A PEACEFUL JIHAD

CONTEMPORARY ANTHROPOLOGY OF RELIGION
A series published with the Society for the Anthropology of Religion

Robert Hefner, Series Editor
Boston University

Published by Palgrave Macmillan

A Peaceful Jihad

Negotiating Identity and Modernity in Muslim Java

Ronald Lukens-Bull

First published in 2005 by
PALGRAVE MACMILLAN™
175 Fifth Avenue, New York, N.Y. 10010 and
Houndmills, Basingstoke, Hampshire, England RG21 6XS
Companies and representatives throughout the world.

PALGRAVE MACMILLAN is the global academic imprint of the Palgrave Macmillan division of St. Martin's Press, LLC and of Palgrave Macmillan Ltd. Macmillan® is a registered trademark in the United States, United Kingdom and other countries. Palgrave is a registered trademark in the European Union and other countries.

ISBN 1–4039–6658–3
ISBN 1–4039–6660–5

Library of Congress Cataloging-in-Publication Data

Lukens-Bull, Ronald.
    A peaceful jihad : negotiating identity and modernity in Muslim Java / Ronald Lukens-Bull.
        p. cm.—(Contemporary anthropology of religion)
    Includes bibliographical references and index.
    ISBN 1–4039–6658–3
    ISBN 1–4039–6660–5 (pbk : alk. paper)
        1. Islamic religious education—Indonesia—Java. 2. Islamic modernism—Indonesia—Java. 3. Islam—Social aspects—Indonesia—Java. 4. Islam and state—Indonesia—Java. 5. Java (Indonesia)—Religion. I. Title. II. Series.

BP43.I5L84 2005
297'.09598'2—dc22   2004043161

A catalogue record for this book is available from the British Library.

Design by Newgen Imaging Systems (P) Ltd., Chennai, India.

First edition: May 2005

10 9 8 7 6 5 4 3 2 1

Printed in the United States of America.

For Katryne, Emmy, and Ronan

Contents

List of Illustrations

Photo

Figures

Table

Acknowledgments

No author works completely alone, nor do they live in a vacuum. So many people have contributed to the completion of this project that a few short words seem inadequate to express my debt of gratitude.

The research for this book was supported by a number of sources including the Program for Southeast Asian Studies and the Department of Anthropology at Arizona State University, the Henry Luce Foundation, and the University of North Florida. I am most grateful to the guidance and direction given me by Mark Woodward, James Eder, John Aguilar, John Chance, James Rush, and Scott Frey. I also wish to thank others who have helped sharpen my thoughts: Brad Biglow, Sharon Estee, Abdurrahman Mas'ud, Rick Phillips, David Ray, and Hermawan Sulistiyo.

In Indonesia, I must thank the Indonesian Academy of Sciences (LIPI) for their assistance in processing the mountains of paper necessary for research in Indonesia. I also thank IAIN Sunan Ampel in Surabaya for being my institutional sponsor in Indonesia; special thanks go to Jabar Adlan, the Acting Rector for taking a deep personal interest in my research; his assistance was invaluable. I am indebted to Abdurrahman Wahid, Yusuf Hasyim, Hasyim Muzadi, Yusuf Muhammad, Badruddin Anwar, Tajab, and Siti Mahmudah. To them I add the men and women whose names I cannot mention.

I thank Bethany Biruk, Heather Frost, Michelle Paul, and Toni Stevens, at the University of North Florida for their tireless and cheerful efforts during the last push: editing, checking the bibliography, and finding last minute sources. I thank David Wilson of UNF's Center for Instructional and Research Technology for rendering the diagrams. I thank Naima Brown for her help in compiling the index.

I thank my wife Katryne, for her support, insights, and wisdom. She has always been my best critic and first editor. I cannot imagine this project or this life without her. I also wish to thank my children Emmelia Marie and Ronan Isaac for teaching me about the truly important things in life.

A Note on Transliteration and Spelling

Indonesian and Javanese words are spelled according to the official convention of the Republic of Indonesia set in 1972. The major changes were dj = j (as in John); j = y (as in yes); tj = c (as in choke); oe = u. The only exceptions to this are words within quotes, titles of books published before 1972, and the proper names of authors and major figures. Arabic words will be spelled as they are in Indonesian except when the words have become common in English. For example: Qur'an, hadith, and shariah.

Glossary

Adat	Custom
Ahklak	Morals; Ethics
ahlus sunna wal jamaah	The Community of the Way of the Prophet
akidah	Theology
Al-Fatiha	First Chapter of the Qu'ran
amal	Good Works; Alms
Asyari	Founder of the Asyari Theological School
barakah	A Spiritual Essence that Conveys Blessing and Power
Bida	Innovation in Religious Matters; Not Allowable
Dakwah	Propagation of the Faith
Doa	Intercession
fikih	Islamic Jurisprudence
Ghazali	Eleventh Century CE Islamic Thinker, Thought by Some to be the Grand Renewer of Islam
hadith	Traditional Sayings and Actions of the Prophet
Haji	Pilgrim
hajj	Pilgrimage
haliqoh	Religious Seminar from the Arabic Halqa
halqa	Study Circle
haul	Memorial Service; Usually for an Islamic Leader
Hukum	Law
ibadah	Ritual Devotions
ijtihad	Interpretation
ikhlas	Selflessness

Imam	Prayer Leader
jamaah	Community; Assembly
jihad	Struggle in the Cause of God
kabah	The Square Black Building in Mecca Toward which Muslims Pray.
kafir	Unbeliever
Khalifa	Deputy (In a Religious Order)
kiblat	The Direction of Prayer
kitab	Book (Generally Used for Classical Religious Texts)
kyai	Javanese Islamic Leader who Combines Roles of both Islamic Scholar and Mystic.
madhab	One of the Four Sunni Schools of Jurisprudence
madrasa	Pesantren-Like School of Islamic Learning, Found Throughout the Islamic World
madrasah	In Indonesia, Religious Schools, Which if not Qualified by Terms Salafiyah or Dinniyah are Part of the National Madrasah System, Which is now Mostly Secular
manakib	Reading the Hagiography of a Saint, Followed by a Ritual Meal
Masjid Jami	Friday Mosque
Maulid	Birth Day of a Prophet
mihrab	Niche in Front of Mosque from Which Prayers are Lead
Mubalig	Preacher
Muhammadiyah	Second Largest Islamic Organization in Indonesia
Muharram	First Month of the Islamic year
Muqadima	Preamble
musyawarah	Consensus Building Group
Nahdlatul Ulama (NU)	"Renaissance of Islamic Scholars"; Largest Islamic Organization in Indonesia and Perhaps the World
nahwu	Arabic Grammar
Niat	Intention
pahala	Merit for Certain Pious Actions
pondok	Literally, Hut. In Religious Discourse can Mean a Place Where Mysticism is

	Studied, from the Arabic *Funduq*. Can also Refer to a Pesantren in the Whole or Just the Dormitories in a Pesantren.
Ramadan	The Fasting Month
Salaf	Worship, Prayers
santri	Student at a Pesantren. Alternatively, Orthodox Muslims in Indonesia.
Shafi'i	Founder of Madhab Followed in Indonesia; Also the Madhab Itself
Shahada	Confession of Faith: "there is one God and Muhammad is His prophet"
shariah	Islamic Law
shorof	Arabic Grammar
Sufi	Islamic Mysticism and Mystics
syehk	Leader of Sufi Brotherhood.
Syirk	Polytheism; Associating other with God
Tafsir	Interpretation (of Qur'an)
tarekat	Sufi Brotherhood
tasawuf	Islamic Mysticism
tasbih	Prayer Beads
tauhid	Oneness of God; Theology
ulama	Islamic Scholar(s)
umat	Religious Community
ustadh	Lower Order Religious Teacher
Usuladdin	History of Islam
wajib	Required
wakaf	Religious Property Trusts.
wirid	Spells Using Qur'anic Text
zakat	"Tithe"; Almsgiving
ziarah	Lesser Pilgrimage to Graves of Important People
zikir	Phrase Chanted to "Remember" Allah

Chapter 1

Negotiating Tradition, Modernity, and Identity

One evening in mid-August 1995, I was working in my rented room when my landlord told me I had visitors: four college students from some of the many colleges in Malang, East Java. One was studying business, another law, the third economics, and the last agribusiness. All four were supplementing their university studies by studying religion at an Islamic boarding school. Although they wanted to participate in the globalization of Indonesia, they wanted their lives to be grounded in a religious lifestyle.

The Islamic boarding school (*pesantren*) at which they studied was Al-Hikam, which in 1994–1995 housed its 60 students in a complex consisting of two two-story buildings and a mosque. Students at Al-Hikam are enrolled in any one of the several colleges and universities in Malang. They can major in any nonreligious field, which, by definition, excludes studying at the *Institut Agama Islam Negeri* (IAIN; State Islamic Institute)—an institute that offers only religiously oriented majors. Demographically and sociologically the students do not significantly differ from other Indonesian college students. Their majors include English, law, economics, business, accounting, political science, and agronomy. On top of their university studies, which the *pesantren* staff monitors, the students have a regular routine of standard *pesantren* education. Some have specific hopes and dreams, but most are eager to find any gainful employment or to successfully launch their own businesses. The only real common denominator is that they do not have any *pesantren* training before entering—and because of the goals of the institution, this is an entry requirement. Students have a number of reasons for attending and living at Al-Hikam while attending college. The most common include wanting to absorb the *barakah* (blessing, but with an existential quality) of the headmaster and to learn more about their religion and the right ways to practice it. In this regard, they differ little from students at other *pesantren*. Al-Hikam students also mention the relatively low housing cost and the access to additional educational opportunities, specifically learning Arabic and English, as additional reasons for enrolling there.

The lifestyle at Al-Hikam differs from more traditional *pesantren*: the students are allowed to watch television and go to the movies in their free time; they sleep on beds, and even eat meat with their meals. In more traditional *pesantren*, as part of ascetic training, sleeping on beds and eating meat are avoided. The watching of television and movies is proscribed because of the morals they portray. At Al-Hikam, the allowance of meat with meals (several times a week, but not nightly) is meant to create a feeling of indebtedness (*hutang budi*) to the headmaster, as to a father; students are said to be more easily taught when they feel indebted. Although beds are used, the mattresses are the thinnest available, and so the living conditions are more ascetic than those in a college dorm, even if they are less ascetic than those in a traditional *pesantren*.

My visitors had sought me out that evening because an exciting event was about to take place at Al-Hikam. Syehk Abdurrahman, a seventh-generation descendant of Sunan Kalijaga, the most famous of the saints who brought Islam to Java and the *syehk* (master, head) of the Qadiri-Naksibandia Sufi brotherhood, was going to visit Al-Hikam early the next morning. This was an important event because it was the meeting of tradition—in the person of Syehk Abdurrahman, who was playing the archetypal role of a Sufi saint—and modernity— in the persons of the college students who wanted to become lawyers, economists, government workers, and businessmen.

Because I was trying to type up field notes, we decided that one of them would pick me up just before the event was to start. At 7 A.M., my ride told me that I should bring my video camera. However, he warned me to ask permission before taping, because Syehk Abdurrahman had such strong spiritual power, coming from his *barakah*, that if I did not ask permission, his image would not show on the videotape. My escort was making a claim that his traditions were more powerful than my modern technology.

When we arrived at Al-Hikam, the gateway to the *pesantren* campus had a bamboo archway built over it. Signs on this structure proclaimed 50 years of Indonesian independence. While awaiting the public appearance of the *syehk*, I sat on the steps of the *pesantren*'s mosque and conducted interviews as I sipped strong coffee while recorded Egyptian music played in the background. These markers place this event into three wider social arenas: Indonesia, by means of the signs on the

Figure 1.1   Photo of Syehk Abdurrahman at Al-Hikam

archway, and both the pan-Islamic world and global consumerism, by means of the recorded chants.

When the *syehk* finally arrived, we all entered the mosque. The *syehk* wore a simple sarong, a white long-sleeved shirt, a green *sorban* (prayer shawl), and an East Java style, or Demak, *peci* (cap). The entire outfit invoked various legitimizing symbols (see figure 1.1).

By themselves, the sarong and *peci* may not be specifically Islamic, but a sarong combined with a prayer shawl (*sorban*) and a *peci* is the special uniform of *kyai*, who are both the headmasters of *pesantren* and the leaders of the Indonesian Islamic community more generally. Once, Yusuf Muhammad, a nationally known *kyai*, told me the story of a con man who wore this outfit to convince people of his holiness and sincerity; to wear this outfit is to represent oneself as a holy man. The style of Syehk Abdurrahman's *peci* was also making a claim. The East Java or Demak style differs from the flat-topped national headgear in that it has a raised crown, though exactly why this is associated with Demak, an early Javanese Islamic kingdom, is unclear. Since the Demak mosque is affiliated with Sunan Kalijaga, the cap is a visual reminder of the *syehk*'s claim of descent from one of the Wali Songo, the nine saints who brought Islam to Java: Sunan Kalijaga is the most famous of the Wali Songo (Florida 1995:28; Woodward 1989:96). Further, Syehk Abdurrahman's prayer shawl was green, the color of the Prophet Muhammad. This color is often associated with descendants of the Prophet, and in some parts of the Islamic world its use is their proprietary right (cf. Gilsenan 1982:9–11; Mottahedeh 1985:30). Indeed, by claiming descent from Kalijaga, Syehk Abdurrahman is also claiming descent from the Prophet.

The mosque was divided into two sections: a men's section, which occupied about two-thirds of the space, and a women's section. There were about 70–80 men and fewer than 20 women. Of the men around 30–40 were Al-Hikam *santri*; the rest were older, in their fifties and sixties. The *syehk* was seated on the floor in front of the mosque next to the *mihrab* (the niche from which prayers are led), flanked by the headmaster of the *pesantren* and a member of his entourage, who in addition to being a *khalifah* (deputy) in the order, is both the regional sales representative for an international tobacco company and one of the three cofounders of Al-Hikam.

Syehk Abdurrahman is said to know the hearts of people; without being told he knows their problems and can give solutions. One by one, the men came up to him and greeted him. For many of them, he had personal messages based on esoteric knowledge. He rebuked some of them, gave messages of hope to some, and prayed with others. When almost everyone else was finished I was told to go forward and greet him. I had lost my *peci* the day before and therefore did not have one with me; so I borrowed someone's skullcap so that my head was properly covered when I greeted him. I crawled forward on my hands and knees, keeping my head lower than his the whole time, as everyone else had. I also followed the example of all the other men and took his hand and kissed it. Unlike everyone else he told me to sit up properly and abandon my almost painful position of supplication and submission. He then announced that there was still a bit of Christian teaching left in me and took me by the right shoulder, where it seemed to be located. He then led the congregation in repeating for me the *Al-Fatiha* (the opening chapter of the Qur'an, used as a prayer whenever Muslims gather) so that such teachings would leave me. He also said that my connections with my parents were too strong and took me by the back of the neck and led the prayer again. Then he took me by the head and told the congregation that they were going to repeat the *Al-Fatiha* three more times so that I would truly believe in Islam, all my doubts would disappear, and I would be able to bring my family to Islam. After my encounter with him, the women (who had not approached him) passed forward some water for him to pray over, so that they could use it for healing the ill. Finally, he led the congregation in reading the proprietary *zikir* (or chant, often repeating the names of Allah) of the Qadiri-Naksibandia Order.

This visit was the final event of two weeks of special training for the students at Al-Hikam. Everyday at *Ashar* (mid-afternoon prayers) for the previous two weeks, they had done *zikir*, that is, chanted Arabic texts in order to "remember Allah" and to achieve mystical states. *Zikir* are passed from master to student, and many Sufi brotherhoods (*tarekat*) have proprietary *zikir*. Every Wednesday night, the Al-Hikam students perform *istighosa* (special intercessory prayers for themselves, the *pesantren*, and their families) lasting for several hours; the form that they use is based on the teachings and practices of the Qadiri-Naksibandia Brotherhood. At the end of the event, copies of the Qadiri-Naksibandia proprietary *zikir* were distributed as were posters depicting Syehk Abdurrahman's genealogy.

After the event was over, the students were nearly ecstatic, especially over my interchange with the *syehk*. They said that they would have loved to have been touched on the head by him, but that even Hasyim Muzadi,[1] the director of Al-Hikam, had only been taken by the hand. They said that I had received a great honor and that I would have much *barakah*. Further, the skullcap that I had borrowed was full of *barakah* because Syehk Abdurrahman had touched it. As a group, they insisted that I keep the cap, against the objections of the owner, and said that they would talk to the owner about being *ikhlas* (selfless, sincere). They said that if I wore the cap to my dissertation defense, I would surely pass; thus, they deftly connected university educational practices and mystical ones.

Syehk Abdurrahman's appearance at Al-Hikam was the result of several factors. Hasyim Muzadi (the founder of Al-Hikam), with the aid of one of his fellow cofounders, who was also a *khalifah* of the *syehk* invited him to visit Al-Hikam. It should be noted that in private meetings during the visit, Hasyim Muzadi was being prepared to be inducted as a *khalifah* of the *syehk* during Ramadan the next year. Hasyim Muzadi explained the reasons for this visit and his subsequent induction as a deputy; the *santri* needed a foundation in Sufism, and Sufi rituals were "concentration practice" (*latihan konsentrasi*) for them. One of the *santri* gave a similar explanation. He said that the reason they needed the Sufi teachings was to balance the education they were getting on campus. It is through such teachings that they learn morality and also gain an anchor against an unsteady soul and psychological problems.

This brief narrative raises more questions than it answers. In the present context, it serves to illustrate a complex process presently taking place in East Java. It shows how a group of people are trying to construct an identity for the Indonesian Islamic community that is both modern and traditional; that is to say, that they want the success and power that come from modernity (i.e., college education) and the benefits that come from tradition, including *barakah* and *pahala* (heavenly reward). I wish to place this event within the wider discourses about identity, modernity, and tradition of which it was part.

Civilizations May Clash; People Will Negotiate

In recent years, the notion of an inevitable clash of civilizations (Huntington 1993, 1996) between the West and the Islamic world has gained considerable play amongst scholars, journalists, policymakers, and even the general populace. In the popular conception, a driving force in this clash is the Islamic principle of jihad, often glossed as "holy war." Neither such an alarmist approach to Islam nor a naive picture of it as entirely at peace with the Western world allows us to fully comprehend the nuances of the interaction between Islam and a globalizing Western culture.

It is true that many Muslims are engaged in a jihad vis-à-vis Western culture. However, as understood by the majority of Muslims, jihad is not holy war but rather "struggle in the cause of God." Jihad is conceived of in terms of a greater and a lesser jihad. The greater jihad is the struggle against one's own sinfulness. The lesser jihad involves public effort. It is in the lesser jihad that one finds the historical wars and the justifications for other acts of violence. For the community discussed here—the Classicalist Muslim community of Java, Indonesia—even the lesser jihad is best done through the peaceful means of education, preaching, community development, and leadership. Hence, I will examine the ways in which this community seeks to engage in jihad, that is, a struggle in the cause of God, to negotiate global cultural changes.

Clifford Geertz, when writing about *pesantren* and their *kyai* 40 years ago, predicted that they would not be able to meet the educational needs of a modernizing country. In general, he was not optimistic about the ability of *kyai* to be cultural brokers between Indonesia and modernity (Geertz 1960b). Not only have *kyai* contradicted Geertz's expectations, but what they are engaging in is not mere brokerage. They are not just translating modernity to Indonesia, but are entering into a processes of negotiation in which they must (re)invent modernity so that it is both Indonesian and Islamic. A key component of this is creating a modernity that is subject to traditional morality. This requires that they must also (re)invent tradition so that it can be used to temper modernity.

Imagining Modernity and Tradition

Ever since the watershed works of the nonanthropologists Eric Hobsbawm and Terrance Ranger (1983) and Benedict Anderson

(1991), invented traditions and imagined communities have become staples of anthropological theory. I suggest that modernity is a trans-local tradition that is invented in the global movement of goods and symbols and (re)invented in local and national settings. Both modernity and tradition are negotiated in terms of each other. As the Javanese and wider Indonesian Classicalist Muslim community seeks to redefine modernity and recreate it in the community's image, so it is forced to (re)invent tradition.

It is a mainstay in anthropology to suggest that tradition is imagined and invented. I wish to add that it is also (re)invented. By (re)invention, I mean the ongoing process by which tradition is invented and reinvented to meet new social needs and challenges. A number of theorists have argued that tradition, and how it is imagined and (re)invented, is inherently political (Briggs 1996; Hobsbawm and Ranger 1983; Schulz 1997; Singh 1998). I would add that modernity is also subject to being imagined and (re)invented and that such processes are also inherently political (Appadurai 1996:3; Freidman 1992:365; Giddens 1990). I take the reflexivity that Anthony Giddens (1990) sees as central to modernity and combine it with Habermas's insistence that modernity must "create its normativity out of itself" (1987:7). I add to this basic orientation Arjun Appadurai's (1996) notion that modernity is not a monolithic whole, but is unevenly experienced, and Jonathan Friedman's (1992) observation that globalization is ultimately experienced and reinterpreted locally. I, therefore, conclude that modernity is imagined and (re)invented in local, national, and transnational contexts. Further, the (re)invention of modernity necessitates the reexamination of tradition. The imagining and (re)invention of tradition and of modernity are two sides of the same coin.

In Indonesian discourse, modernity and tradition mean a great number of things; these are terms oft used and seldom defined. Modernity can mean advanced technology, such as airplanes, computers, and automobiles. It can mean consumer goods and global popular culture. It can also mean simply nontraditional. Tradition is likewise ill defined if oft debated. It can refer to Javanese culture, to the early centuries of Islam, or to the accumulated wisdom of the *madhab*.

The first Western scholarly attempts to define modernity came from modernization theories that were, at core, concerned with the transformation of a society's institutions. One concern of modernization theorists was economic modernization, which included the application of scientific technology, the specialization of labor, and banking and finance. These modern institutions would raise the nation's level of material well-being (Ward and Rustow 1964). Another concern of modernization theorists was political modernization, which was thought to

involve three processes. The first is the replacement of traditional institutions of authority (religious, familial, and ethnic) by secular, national
political authorities. The second is development of new administrative
hierarchies to manage new political functions. The third is the development of new institutions such as political parties and interest groups
to organize increased participation in politics by social groups
throughout society (Huntington 1966). Few, if any, of these institutional changes have ever occurred in a nonproblematic fashion. Further,
modernization theorists have been rightly criticized for being Euro-
centric. However, their basic ideas about modernity continue to influence how some scholars and some Indonesians continue to think about
modernity.

In his consideration of the modern movement of commodities and
images, Appadurai argues,

> Globalization does not necessarily or even frequently imply homog-
> enization or Americanization and to the extent that different societies
> appropriate the materials of modernity differently, there is still ample
> room for the deep study of specific geographies, histories, and languages.
> (1996:17)

Hence, my major contribution is to explore the specificities of how
the Indonesian Classicalist Muslim community appropriates the materials of modernity. In this appropriation, the leaders of this community
are concerned with the deleterious effects of modernization, as they
see them—egotism, materialism, social inequities. Further, I explore
how, despite Appadurai's claim above, these leaders see the negative
aspects of modernity as essentially the Western, if not American, trim-
mings on the house of modernity. As part of their appropriation of the
materials of modernity and their subsequent reinvention of modernity,
these leaders have created an educational system to address both the
educational needs of a modernizing society as well as guard it against
perceived moral decay. The influences of global popular culture and
capitalism are seen as morally threatening and so these leaders created
an educational environment in which the character development of the
students is a planned and carefully considered part of the curriculum.

However, it is clear that there are elements of modernity that may
not be discursive. The movement of global finances and global capitalism is largely beyond the control of local actors. Whether Indonesians
like it or not, Indonesia is part of global capitalism, and for that matter placed firmly in the periphery as a source of raw materials (oil,
hardwoods, ore) and of inexpensive manufactured goods. In fact, as
companies shy away from Indonesia due to recent terrorist threats,

Indonesia's economy is delivered another blow. And so, the government seeks to attract new business. However, although certain elements of modernity are not discursive per se, people can and do evaluate their impact. Through discourse, people decide what opinion they will hold about the more intractable dimensions of modernity.

Education is widely recognized as an important part of how a society engages modernization and globalization (Demerath 1999; Li 1999; Wood 1976:49). The Indonesian Classicalist Muslim community has used education to appropriate the materials of modernity and subsequently (re)invent them. While *pesantren* seem to have been of passing interest to Western scholars (Anderson 1990:64–65, 127–128; Geertz 1960a:180–187, 1960b; Jones 1991), Indonesian scholars have produced an enormous literature on them, including countless books and scholarly theses. Most of this literature is firmly based on the work of Zamakhsyari Dhofier (1980b, 1982) and Taufik Abdullah (1987), which remain good introductions to the study of these schools. A large number of these works assert that *pesantren* and modernity are not incompatible but can work together for the betterment of the nation (see especially, Galba 1991; Prasodjo et al. 1974). The leaders of this community have created an educational system to address the educational needs of a modernizing society and, at the same time, to guard against the perceived moral decay that comes with modernization and globalization. I draw upon Eleanor Leacock's (1976) observation that how modernity is imagined shapes how educational systems engage and negotiate it.

Imagining Modernity in Order to Redefine it

Before the leadership at Al-Hikam and other Indonesian Muslims can engage in the (re)invention of modernity and tradition, they must imagine a modernity in need of revamping. They also need to imagine a tradition that is able to fit with and temper modernity. By imagining, I mean having a particular understanding of something and the ability to see it somehow transformed. The imaginings and (re)inventions considered here are first part of *pesantren* discourse. However, because of the importance of *kyai* and *pesantren* to the majority of Indonesian Muslims, these imaginings quickly become part of public discourse.

Imagining Modernity as Dangerous

In the act of imagining and (re)inventing modernity, *pesantren* people first imagine it as potentially dangerous. In this regard, many *pesantren* people associate the processes of modernization and globalization with the loss of traditional values. One teacher at Al-Hikam lamented that Indonesia had lost its own value system:

> Indonesia once had established values, as can be seen in the successful establishment of the Republic of Indonesia. These values were the values of 1945. However in the 1980s, these values began to be lost and are now completely lost. The problem is that these days, young people want to be like the United States or Japan as quickly as possible. But they often forget that Japan has held on to its values tenaciously. The Qur'an can guarantee life in the future; the Qur'an can take us back to the values of 1945.

The values of 1945 are debated in Indonesian discourse, but many in the *pesantren* world hold them to be uniquely Islamic. They point to the Jakarta Charter that became the preamble of the constitution. The original version required all Muslims to uphold Islamic law. Further, the establishment of the republic and the War for Independence (1945–1949) are typically described in *pesantren* circles as being based on the *pesantren* values of self-sufficiency, self-sacrifice, and healthy distrust of any ruler, particularly a non-Muslim.

American films and television are central components of a dangerous modernity. In the early 1990s, the United States required Indonesia to import American films and television shows in order to continue to export textiles to the United States (Barber 1995:91). Repeatedly I heard concerns from *pesantren* people about the American movie industry's purported intention of destroying Islam and corrupting the values of Islamic societies such as Indonesia. Many were concerned with the portrayal of scantily clad women (having bare shoulders and knees). Such concerns persist even though Appadurai asserts, "the United States is no longer the puppeteer of a world system of images but is only one node of a complex transnational construction of imaginary landscapes" (1996:31). I have heard other *pesantren* people express concerns with what they see as the deliberate imitation of the values portrayed in the television show, *Beverly Hills 90210*, and other American programs by blue-jeans wearing, disco attending, alcohol drinking youths. These concerns show how *pesantren* people imagine a modernity that is threatening and in need of (re)invention.

How the leadership of Al-Hikam is imagining modernity is further seen in the basic mission of the school. Al-Hikam was designed to mold modern scholars who have strong religious values. Al-Hikam also seeks to create a balance between general education and religious education for university students. However, most of Al-Hikam's curriculum is religious; the general education comes from the students' university studies. There is a clear sense that science, as part of modernity, lacks the moral imperatives to guide society. Or as Hasyim Muzadi has stated, "Colleges create smart people, but it is uncertain whether they are moral people." Even in the West, there is the axiom that states that science asks if something can be done but rarely asks whether it should be done. Hence, modernity is, at best, amoral. Certainly some of the technological trappings of modernity (i.e., airplanes, computers, the internet, television, and so forth) are morally neutral in and of themselves. On the other hand, many of the trappings of globalization, especially global popular culture, are seen as promoting a blind, self-interested consumerism that threatens to unhinge society.

Imagining Modernity as Malleable

In addition to seeing modernity as potentially threatening, imagining modernity also involves conceptualizing modernity as something that can be (re)invented. One basic way *pesantren* people imagine a co optable modernity is found in an oft-quoted Arabic principle that says that one should continue in old ways that are good and adopt new ways that are better.[2] This phrase empowers *pesantren* people to (re)invent modernity. Nafik, Hasyim Muzadi's assistant said modernity is not the external trappings, but is rather a "frame of thinking," an opinion shared by Robert Bellah, who pointed out that modernity should be seen not "as a form of political or economic system, but as a spiritual phenomena or a kind of mentality" (1968:39). This is precisely the component of modernity with which *pesantren* people are most concerned.

By shifting the definition of modernity to "a frame of thinking," the leadership at Al-Hikam, and other *pesantren* leaders, have placed it in the same realm as religion and hence they are able to insert certain values and morals in their conceptualization of modernity. These values include Islamic brotherhood (A., *Ahwuya Islamiya*), selflessness (I., *keikhlasan*), simplicity in living (I., *kesederhanaan*), and self-sufficiency (I., *kemandirian*). Also included is a concern for social justice and serving the needs of the poor. Taken together, these values define a modernity quite different from that generally practiced in the West. Perhaps this can avoid or overcome the tendency, identified by Mitchell (1976), for educational systems in developing countries to both create new hierarchies and to reify old ones that serve the interests of national and transnational elites.

The Ethnographic Setting

Indonesia is the world's largest Islamic country, although it is not an Islamic state. Throughout the Indonesian Republic's existence, the ongoing question for the Islamic community has been how to create a strong, pious, and faithful Islamic society in the context of a modernizing, globalizing, and secular state (Abdullah 1996:65; Boland 1971:15–34; Hefner 2000; Horikoshi 1975:60; Noer 1978:12). Through developing a hybrid educational system in *pesantren*, *kyai* have outwardly supported the national development policies while striving to firmly establish Islamic values as the foundation for public life in Indonesia.

Earlier Anthropological Views of Islam in Indonesia

Until recently, the dominant paradigm for the study of Java has been subtly Orientalist and anti-Islamic. It operated as if the only Islam is that of "fundamentalists," which were often little more that shadowy specters. There was a tendency to distrust, dislike, or merely downplay Islam (Anderson 1972; Geertz 1960a; Kahin 1952). Other works assert that Islam is not significant in the religious, social, or political spheres of Javanese life (e.g., Benda 1958; Kahin 1952:43).

Geertz suggests that, in Java, there are three main social–structural nuclei: the village, the market, and the government bureaucracy. These nuclei connect Geertz's five occupational types: "farmer, petty trader, independent artisan, manual laborer, and white-collar clerk, teacher, or administrator" to his three cultural/religious types: *abangan*, *santri*, and *priyayi* (1960a:4). In this schema, the nominally Muslim peasant villagers adhere to the *abangan* tradition, which forms "a basic Javanese syncretism which is the island's true folk tradition" (Geertz 1960a:5). *Santri* are middle-class traders, village chiefs, and well-to-do peasants whose economic lives center on the market. *Santri* are more conservative in their expression of the Islamic faith than either the *abangan*, or the *priyayi* (1960a:5–6, 40–41). The *priyayi* are white-collar government bureaucrats descended from the traditional aristocracy (1960a:6). They practice a form of religion derived from the court Hindu–Buddhism of the pre-Islamic era. Woodward argues that the Geertz's view of Javanese religion became paradigmatic and "is exemplified by otherwise credible scholars taking the marginality of Islam in Indonesian culture as a given" (1996a:31).

Shortly after the publication of Geertz's ideas about the nature of religion and society in Java, the *abangan*, *santri*, *priyayi* trichotomy became a central concept in Java studies. It was a useful handle to deal with the complex religious and social reality of Java (cf. Benda 1958; Lyon 1970:29ff). The paradigmatic nature of Geertz's analysis cannot be attributed to his efforts of Geertz. I agree with White that readers of Geertz overenthusiastically took his inspired speculation as a "discovery" (White 1983:22). Stange agrees that Geertz is misread and asks whether authors can be blamed for the misuses of their work (1990:241–242); if the work lends itself easily to such misuse, perhaps. However, this is a moot point. In order to abandon the paradigm we must show it to be lacking, and the most effective way to do so is to challenge its foundations. Any work, whether misread or not, which forms the foundation of a paradigm is subject to such scrutiny.

Another variation of the paradigm suggests that the conversion of Java was never more than a surface change. Benda argues that Indonesian Islam began as a largely urban phenomenon (1958:10). In those areas where Hindu civilization had been strong (e.g., Central and East Java), Islam did not have a strong impact upon the religious, social, and political spheres (Benda 1958:12). In these areas, Islam adapted itself to the part-Javanese, part-Hindu–Buddhist traditions that preceded it. According to Benda, the greater significance of Javanese Islam was in politics rather than religious affairs; a change of faith to Islam "did not bring about radical change in religious and social life on Java" (Benda 1958:12). Such a position is echoed by M.C. Ricklefs, who denies that the conversion to Islam significantly altered the fundamentally Hindu/Buddhist character of Javanese religious thought,

> To be Javanese is, for the majority, to be abangan Javanese; the santri Javanese is perceived by the bulk of Javanese society as person who has to some extent removed himself from the social and cultural environment. (1979:127)

This variation of the paradigm assumes that to be a "real Muslim" one cannot be a "real Javanese" (Woodward 1996a:33). Paradoxically, while Benda recognizes East Java as a Hindu stronghold, others recognize it as where the most advanced *pesantren* in Java are found (Kumar 1985:11). Interestingly, both claims are true and the conflict between these two social realities has been expertly examined by Robert Hefner (1985, 1987).

The paradigm in Java studies that states that Islam is not important to the Javanese is indicative of a general problem in the anthropological study of Islam. Javanists took an essentialist view of Islam that defined it according to a *syariah*-centric orientation. Local Islamic practices fell outside of this limited definition of Islam and were declared non-Islamic. However, there is great diversity in what practicing Muslims define as Islamic. Scholars have now turned their attention to the discursive tension between the universal aspects of Islam and the local situation (Bowen 1993b:7). One locus of this discursive tension is in education; the question is how to teach people to be good Muslims given the existing social and political conditions (Abdullah 1987; Dhofier 1980b).

Recent scholars like Woodward, Hefner, and Bowen have begun to correct this paradigm. Woodward is probably Geertz's strongest critic. His work suggests that a broader understanding of what is Islam is necessary. Anthony Johns would concur that the term "Islam" has often been used with little precision. There is great variation within Islam, even within what might be called "orthodox Islam." This necessitates an understanding of these variations as they interacted with various Southeast Asian local contexts. Dale Eickelman argues that the primary task of scholars of Islam in local contexts should be to understand "how the universalistic principles of Islam have been realized in various social and historical contexts" without either seeing Islam as a "seamless essence" or a mere aggregation of beliefs and practices (1982:1).

Variants of Islam in Indonesia

There are two major variants of Sunni[3] Islam in Indonesia, which I will refer to here as Classicalist and Reformist, sometimes referred to as Traditionalist and Modernist. Classicalists are typified by their use of classical Islamic texts and their affiliation with *pesantren* and the organization Nahdlatul Ulama (NU, Renaissance of Islamic Scholars). Reformists, who are affiliated with the organization Muhammadiyah, seek to reform Indonesian Islam so that it draws primarily on scriptural sources (Peacock 1978). Two smaller groups need to be briefly mentioned. There are some people who seek to bridge these two approaches and are referred to as Neo-Modernists or Neo-Traditionalists. Finally, in recent years, especially since the fall of the Suharto regime, Islamist groups have emerged in Indonesia.

The Classicalist variant is centered around *pesantren* and their headmasters (*kyai*), are hence the leaders of this religious community. The terms "*pesantren* world" (*dunia pesantren*) and "*pesantren* people" (*orang pesantren*) are preferred by most members to the exonym "traditionalist *santri*," a designation made popular by Clifford Geertz (1960a). When used in the *pesantren* community, the term "*santri*" refers to a student in a *pesantren*, or pe-santri-an, the *santri* place. The *pesantren* community practices and maintains Classical Islam which Zamakhsyari Dhofier sees as

> still strongly bound up with established Islamic ideas created by scholars, jurists, doctors, and Sufis during the early centuries of Islamic theological and legal development, sectarian conflicts, and the rise of Sufi movements

and brotherhoods in the thirteenth century. This is not to say, however, that contemporary classical Islam in Java remains fixed in the molds created for it by the *ulama* (Muslim leaders) of the formative centuries. (1999:xix)

The theologies, considered opinions, legal theories and findings, and mystical theories of Classical Islam are found in texts called *kitab kuning* that is an Arabic–Indonesian hybrid word literally meaning yellow books. However, the term refers to what van Bruinessen has called the classical texts of Islam. It should be made clear that classical refers not to the original Meccan and Medinan communities but roughly to the medieval period, specifically from the twelfth to seventeenth centuries C.E. in which being Muslim and being Sufi were nearly synonymous. The *pesantren* community holds them to be of high importance in determining how to live as good Muslims in a globalizing and modernizing world. They are critical components of *pesantren* curricula.

The Reformist branch of Indonesian Islam, which is affiliated, in part, with Muhammadiyah, has established Islamic schools modeled after the *pesantren*, but none of these are recognized as true *pesantren* by Classicalists. Hence, in the usage of *pesantren* people, contemporary Reformists, can never be true *santri*, although Geertz (1960a) labels them "modernist santri." Muhammadiyah takes a position that the basis of Islamic Law (*syariah*) is the Qur'an, hadith (the sayings and actions attributed to the Prophet), and personal interpretation. They thereby reject historical developments in Islam and Classical Islamic scholarship.

However, it should be noted that the current concern in the *pesantren* world and NU is how to be both modern and faithful to the classical teachings. I wish to avoid the Traditionalist–Modernist dichotomy, which suggests a nonprogressive, staid, and nonrational approach on the part of *pesantren* people and employ the terms Classicalists and Reformists. Classicalists are those who use, read, and study *kitab kuning* (lit., yellow books) that include the writings of Al-Ghazali and other "scholars, jurists, doctors and Sufis of the formative centuries of Islam" (Dhofier 1999:viii). Martin van Bruinessen argues that Reformists are those who use Latin script Indonesian texts, which he refers to as *buku putih* (white books) in contrast to the "yellow books" of the Classicalists (1990:227). However, at higher levels of Muhammadiyah, there may be familiarity with historical texts. Further, by the time of my fieldwork (1994–1995), many Classicalists have published books in Indonesian including Sirajuddin Abbas, whose works are seen as solid summary of *pesantren* world jurisprudence and religious thought

(Federspiel 1996). Abbas' four volume *Forty Religious Problems (Empat Puluh Masalah Agama)* (1995) is used in many *pesantren* as an intermediary reference text and has been reprinted 24 times since it was first printed in the late 1960s and early 1970s.

The common terms *Traditionalist* and *Modernist*, are mostly in reference to the question of *ijtihad*, which is most often translated as interpretation, but may be more correctly defined as "working with the sources of dogma" (Vikør 1995). Classicalists hold that the door of "interpretation" (*pintu ijtihad*) is closed and following the broad outlines of interpretation set forth by earlier great scholars *ijtihad* of the Qur'an and hadith that became the basis of the four schools of jurisprudence (*madhab*). Later historical scholars, such as Imam Al-Ghazali, and even the fourteenth-century father of Islamic reformism,[4] Ibn Tamiya were working as *mujtahid fil madhab* (interpreters of a particular law school) (Cheneb 1953:151; Federspiel 1996:206). Reformists favor the position that the door of interpretation is open and that a properly qualified scholar must have the right to perform *ijtihad* at all times. The Classicalist position (e.g., NU's) does not mean a blind commitment to following established dogma, but means that forming opinions must be done within the frame of previous scholars. Indeed, Abdurrahman Wahid favors a position in which the methods and general framework of previous scholars are used, rather than a strict observance of specific past decisions. In this regard, he states

> However, I am of the opinion that following the Shafi'i *maddhab* means following Imam Shafi'i in his methodology (*manhaj*) for determining law, specifically following his principles of jurisprudence (*usul fiqh*). Therefore, we must continually reformulate what following a *maddhab* means. (in Hamzah and Anam 1989:26, translated from Indonesian)

Although Abdurrahman Wahid is seen by some Western observers as a Neo-Modernist (Schwarz 1994:178), it would be just as accurate to call him a Neo-Classicalist. Although Abdurrahman has "liberal, modernist attitudes" (Woodward 1996b:133), he has repeatedly argued that it is crucial that the teachings of Classical Islam should not be lost. He states that some *kyai* like Hasyim Muzadi of Al-Hikam may not have memorized classical texts but they know the contents and base their teachings upon them. This is the bare minimum for Abdurrahman; there is still a need for those who can and do read and study the texts in Arabic. Abdurrahman himself is conversant in these texts and can quote them extemporaneously. This commitment to classical texts and interpretations is characteristic of NU *ulama* regardless of their position on political, educational, social, and other issues.

Basic Pesantren *Education*

Nahdlatul Ulama, the largest Islamic organization in Indonesia and perhaps the world, was founded in 1926 in Surabaya by several East Java *kyai* including Kyai Haji (KH) Hasyim Asyari of Tebu Ireng, and Kyai Wahab of Tambak Beras. About the relationship between NU and the *pesantren* world it is often said,

> NU is the *pesantren* writ large.
> The *pesantren* is NU writ small.

This slogan affirms the close relationship between the organization and its roots in *pesantren*. This analogy is taken one step further when some described NU, in 1995, as the *pesantren* headed by Kyai Abdurrahman Wahid.

However, Abdurrahman acknowledges that the relationship between NU and *pesantren* is not exclusive. Not all *pesantren*, or *kyai*, belong to NU. Conversely, according to Abdurrahman, the NU chapter in Jombang does not want to be associated with the *pesantren* because of the egoism of the Jombang *kyai*. He argues, further, many *kyai* are involved with organizations other than NU.

Prior to the twentieth century, *pesantren* were the only formal education institutions found in Java and in most of what is now Indonesia. They taught an almost exclusively religious curriculum to a mix of students including future religious leaders, court poets (Florida 1995), and members of the ruling class (Adas 1979; Pemberton 1994:48–49). First the Dutch, then the Nationalists, and later the Republic of Indonesia promoted an educational system that focused on science, math, and other secular subjects (Anderson 1990:132, 243). In response to the demand for this type of education, in as early as the 1930s, many *pesantren* added government recognized curricula. Starting in the 1970s, these new curricula became an important part of the *pesantren* community's strategy for negotiating modernity. These curricular changes can be attributed to a number of factors including competition from secular schools, government grants, and Islamic revitalization. Particular emphasis should be placed on the Islamic resurgence of the 1970s and 1980s that put an emphasis on spreading and strengthening Islam through teaching and preaching. Combined with a loss of cliental to secular schools, which was leading to the decline of the *kyai*'s influence, this emphasis on changing society through education made *pesantren* and other forms of Islamic education central to the Indonesian Islamic efforts to negotiate modernity.

These changes have shaped both the daily round as well as a general sense of what kind of education a *pesantren* should provide (Dhofier

1999). It is common for parents to seek schools that give their children the necessary skills and knowledge to do well in the modern job market and the moral and religious training to be good Muslims and upstanding citizens.

Pesantren exist for children and youth of all ages and at all stages of education—primary, secondary, and tertiary. Although there are *pesantren* for both males and females, they are generally gender segregated. Today, 20–25 percent of Indonesia's primary and secondary school children are educated in *pesantren*-based schools. In some areas, such as Aceh, this number may be as high as 40 percent (Zamakhsyari Dhofier, personal communication, May 1995). Although *pesantren* have their roots in rural setting, *pesantren* today are found in both urban and rural settings and attract students from both urban and rural areas. Although many, if not most, *pesantren* students are from lower socioeconomic levels, *pesantren* still attract middle- and upper-class students.

It is common for *pesantren* to engage modernity by opening government curricula schools at the junior high and high school levels. Contemporary *pesantren* at this level aspire to deliver both the pre-twentieth-century *pesantren* curriculum and the newer government curriculum with the hope that the graduating alumni have the religious knowledge and morality of a religious leader as well as the basic education needed to pursue further education at the college level. The curriculum found in contemporary *pesantren* can be divided into four basic areas: religious education (*ngaji*), character development (*pengalaman*; lit. experience), vocational skills training (*ketrampilan*), and general education (*sekolah*). The first three types of instruction are rigidly gender segregated. At some *pesantren*, general education may be coeducational following the example of the government schools, however, this has proven to be somewhat controversial.

An additional, even central, component of *pesantren* religious education is Islamic Mysticism or Sufism. Key texts are studied and mystical practices, such as *zikir* (chanting religious formulae; lit. remembrance of God) are integrated into daily activities. Sufism as practiced in *pesantren* insists on a mysticism subject to normative Islam and distinguishes intellectual, emotional, and organizational components of Sufism (Dhofier 1999:137, 158). Related practices are special intercessory prayers called *istighosa*, repetitive chants, communal meals to invoke blessing, and grave visitation.

In addition to religious education and character development, many *pesantren* have curricula designed to teach their students the skills and knowledge to find employment after they graduate. General education usually includes one of two basic government recognized curricula, one mostly secular and the other with a greater emphasis on religious training. *Pesantren* may have neither, either, or both types of schools within their grounds. Skills training includes welding, automotive mechanics, furniture carpentry, sewing, computers, shop keeping, and other vocational skills. How exactly *pesantren* at this level accommodate these areas in their curriculum is part of how they negotiate modernization and globalization.

The processes discussed here take place in a wider context filled with many competing voices. Not all the voices are considered here, but the ones examined here speak within a wider context. The other voices in the national process of social construction include: Muhammadiyah and the modernists, who have little, if any, affiliation with *pesantren* (Noer 1973; Peacock 1978), Hindus (Hefner 1987; Lansing 1983, 1995), non-Javanese Muslims (Bowen 1993b), upland horticulturists with an unrecognized[5] minority religion (i.e., animism) (George 1996; Tsing 1993), royal mystics in the Javanese courts (Woodward 1989), Javanese shadow puppet *dalang* (puppeteers) (Keeler 1987), the armed forces (Arto 1994), and the national government (Noer 1978).

Research Settings and Methods

Ethnographic research was conducted in East Java from November 1994 to October 1995, taking a regional rather than a village-based approach. There were return visits in 1997 and 2000 of several weeks each. Although still uncommon in anthropology, regional ethnography is justified when the subject is not limited to a single setting, and others have begun to find it appropriate as well (e.g., Babb 1975). The negotiation of modernity and identity in Indonesia is not limited to a single village; it entails a national discourse.

My taking a regional approach is not to say that this research did not focus a community. It most certainly did, but not on a community bound by a village. In the most general sense, the community I studied was that of all Islamic believers, the community to which those studied see themselves as belonging. They narrow their religious community to *ahlus sunna wal jamaah* (the community of those following the Prophet's example). But since this is a contested term, it may mean little more than "the real Muslims" and may not help us in defining the real group of people with whom my informants identify. At the next lower level, the community is the *dunia pesantren* (*pesantren* world). More narrow still, the community is that of *ulama* (Islamic scholars). These men often meet formally and informally. Formally, there are *haliqoh* (seminars; Arabic, *halqa*, study circle), *Bahtsul Masail* (meetings to determine religious law), *mushawara* (deliberations, often regarding the official structure of NU), *pengajian* (public sermon events to which two or more *ulama* may be invited). Less formally, there are life-crisis rituals for these men and their families. It is not uncommon for *kyai*, which are a kind of *ulama*, from all over East Java, and sometimes from further afield as well, to attend the weddings, circumcisions, memorial services, and other rituals hosted by a *kyai*. It is this regional community of scholars that meets in various locations that justifies a regional study. However, this regional community of *ulama* is also localized. Each *kyai* leads a local religious community centered around a *pesantren*. The degree to which this local community extends beyond the walls of the school varies for each *kyai*. Therefore this regional ethnography focused mostly on three *pesantren* and the communities around them.

A regional approach in ethnography is also the logical outcome of recent trends in anthropological studies of Islam. John Bowen argues that the main impediment to the anthropological study of monotheisms is that these religions do not fit well in the normal ethnographic model. The texts and rituals common to a monotheism transcend any particular locale and take the believer, and should take the ethnographer, outside the village to a "worldwide confessional community" (Bowen 1993a:185). A regional study allows us to explain processes beyond the boundaries of a single village. This is still a limited view and does not encompass the whole Islamic world. This is a limitation of any fieldwork; we need to keep in mind that this identity construction process in Java is part of a larger Islamic discourse about morality, religiosity, and identity.

Interviews were conducted both in the *pesantren* world and with people in the general population ranging from such high ranking figures as Abdurrahman Wahid, then head of Nahdlatul Ulama, and Zamakhsyari Dhofier, then a high ranking official in the Ministry of Religion and the Director of Islamic Schools Development Program to *kyai* of varying degrees of fame and notoriety. I also interviewed faculty members and students at IAIN (State Islamic Institute) campuses, because that system was born of the *pesantren* world. Also included were villagers who lived around the *pesantren*, as well as *santri* from the youngest and newest to the most trusted *ustadh* (senior student and junior teacher). The research also included people outside the *pesantren* world: some reformist Muslims; members of Muhammadiyah, including former Muhammadiyah functionaries; and past and present faculty members and administrators of Universitas Muhammadiyah campuses. The reason for this wide-range interview strategy was to develop a picture of where the *pesantren* milieu fits into wider society.

While this research speaks to wider Javanese and Indonesian society, it was conducted in East Java. The Javanese constitute 40–45 percent of the population of Indonesia. More than 30 million people live in

East Java, with the largest population centers being Surabaya with nearly two and half million residents, and Malang with nearly 700,000 residents. The population of the region averages 672 persons/km^2 with the lowest density in Banyuwangi with 251 persons/km^2. The highest population density is Surabaya with 8,395 persons/km^2. The population density in Malang in 5,774 persons/km^2 (Kantor Statistik Propinsi Jawa Timur 1993). Further, East Java is the recognized center of the *pesantren* world; many prominent leaders of the Islamic community, both Classicalists and Reformists come from East Java. In 1982, there were approximately 4,000 *pesantren* in Indonesia, 1,800 of which were found in East Java (Ghofir et al. 1982:ii). In 2000, researchers at the IAIN (National Islamic Institute) in Semarang estimated there were about 9,000 *pesantren* throughout the country (Abdurrahman Mas'ud, personal communication, August 2000). Estimates still hold that over 40 percent of all *pesantren* are found in East Java. Although centered in East Java, this educational movement has established *pesantren* throughout the archipelago. The *pesantren* selected for research were those that are engaged in the ongoing process of defining and redefining *pesantren* education.

Extended ethnographic research was conducted in three *pesantren*: An Nur in the *Kabupaten* (Regency) of Malang, Tebu Ireng in Jombang, and Al-Hikam in the city of Malang. Shorter visits were paid to other *pesantren*, the data from which I draw on as needed. The schools selected are all engaged in redefining the role and function of *pesantren*, and hence the nature of the *pesantren* world. The three main cases suggest different models for both Indonesian and *pesantren* world identity as well as for *pesantren* curriculum. This ethnographic research established the nature of the daily activities of both *kyai* and *santri*. This included data about the general nature of the educational process including the amount of time spent in various activities. This participant observation was used to establish a baseline from which to compare other *pesantren*. Interviews were conducted with several *kyai* to inquire about the direction of *pesantren* education and what the needs of *santri* will be. After this baseline was established, several shorter visits were paid to various *pesantren* and *kyai*.

Research Sites

Tebu Ireng has about 1,500 students, all male, but is also part of a complex of family *pesantren* that includes *pesantren* for female students, some of whom attend the National System schools in Tebu Ireng. This *pesantren* gives a slight emphasis to "secular" education over traditional *pesantren* education. It has a rich history that is intertwined with that of the Republic of Indonesia; Tebu Ireng's founder, Hasyim Asyari was a cofounder of Nahdlatul Ulama, the largest Islamic organization in Indonesia and several of his descendants including Abdurrahman Wahid (also known as Gus Dur), the current general chairman of NU, have played and continue to play important roles in the Indonesian public sphere. This national range of influence is reflected in the fact that Tebu Ireng's students come from all over Indonesia. Further, the grave of its founder in the heart of the school is an important pilgrimage site that draws several thousand visitors each month. In addition to my own research, a number of Indonesian scholars have written about Tebu Ireng (Arifin 1993; Dhofier 1980b).

An Nur has about 500 students, mostly from East Java, and mostly male (it had 19 female students in 1995). However, it is part of a larger complex of family run *pesantren* that includes another 1,000 students, 300 of which are female. An Nur has a history of less than 50 years and is run by the sons of the founder. Although both secular and religious education is offered at An Nur, the focus is decidedly on the traditional *pesantren* curriculum. The range of influence of An Nur and its *kyai* is limited to area around Malang as evidenced by the fact that nearly half of its students come from this area.

In 1995, Al-Hikam was just three years old and it had 60 male students. It differs from both Tebu Ireng and An Nur in that it did not grow out of a traditional *pesantren*, but was designed as a place where college students can study traditional *pesantren* learning and mysticism while pursuing their college education. All of the students attend college in Malang and most are from East Java. The headmaster, Hasyim Muzadi was the head of NU East Java, and a *khalif* (deputy) in the Qadiri-Naksibandia *tarekat* (Sufi order).

As mentioned above, the schools selected are all engaged in redefining the role and function of *pesantren*, and hence the nature of the *pesantren* world. The three main cases suggest different models for both Indonesian and *pesantren* world identity as well as for *pesantren* curriculum. This ethnographic research established the nature of the daily activities of both *kyai* and *santri*. This included data about the general nature of the educational process including the amount of time spent in various activities. Participant observation was used to establish a baseline from which to compare other *pesantren*. Interviews were conducted with several *kyai* to inquire about the direction of *pesantren* education and what the needs of *santri* will be. After this baseline was established, several shorter visits were paid to various *pesantren* and *kyai*.

Negotiating Modernity and Tradition, Contesting Identity

The debates about modernity and tradition and the efforts to negotiate these processes are essential parts of identity construction for the Classicalist Islamic community in Indonesia. The discursive nature of Islam means that it is concerned with deciding what is allowable (*halal*) and what is not (*haram*). This discourse may be formal and found in seminars, classrooms, and public speeches, or it may be informal and found in coffee shops and living rooms. Anytime people discuss how to be moral they are engaged in moral discourse, which is not merely about right and wrong; it is about identity. What we define as moral, defines us as a people.

James Hunter suggests that discourse about the moral basis of society can be referred to as a "culture war" (Hunter 1991). He developed this notion specifically to discuss the culture war in America that evolved out of century old tensions between competing faiths (1991:67).[6] Extrapolating a wider theory from Hunter, cultural wars may erupt in places where people working with different faiths, religions, or faith-like ideologies attempt to discourse about the nature of society and identity of community, whether local or national. Further, culture wars are simultaneously about morality and identity.

The present work shows how the members of a particular segment of East Javanese society are trying to shape the religious and moral foundations of both East Javanese and wider Indonesian society. Hunter also reminds us of the place of education in any conflict over the nature of society. Although Hunter stresses that education is one of the contested issues, the East Javanese case shows that it is also one of the arenas in which the conflict is fought; through education, young men and women are given a particular view of the world, which they are then encouraged to carry out into society.

What Indonesian Islamic education should be is a hotly debated point. Many fear that the *pesantren* world will face a moral crisis if the moral and religious education in *pesantren* is not guarded. However, the demand for general education has forced many in the *pesantren* world to ask themselves what the future of *pesantren* education is. Will it continue to be exclusively religious and create *ulama*? Or will it be mostly secular, with enough religious training to create a religious and moral workforce? This is the core of the discourse about *pesantren* education.

The process of negotiating modernity and tradition in this community starts with education and so shall this volume. In the next chapter, the book turns to a discussion of how the leadership of one school, which also serves as an important pilgrimage site, have structured the physical landscape of that school to model what they see as the right kind of relationship between modernity and tradition. The third chapter is an overview of the curricular debates and trends. It examines how the *pesantren* community negotiates modernity and tradition through curricular restructuring. In the fourth chapter, we move beyond schools and education to explore the efforts to shape traditional and modern identities in the *pesantren* community. The fifth chapter examines leadership within the *pesantren* community and examines the negotiation and construction of leadership as it relates to creating a community identity that engages both modernity and tradition. It then turns to the context of the presidency of Abdurrahman Wahid, who is both *kyai* and president and examines how dimensions of leadership are part of the process of negotiating modernity and tradition. In the concluding chapter, the discussion moves beyond Java and Indonesia and draws some conclusions that are useful for understanding similar processes elsewhere in the Islamic world.

Chapter 2

The "House" that Change Built

Once, a *kyai* and his senior *santri* came to *pesantren* Tebu Ireng to meditate and pray at the grave of its founder, Hasyim Asyari. The *kyai* was looking for the *tasbih* (prayer beads) of Hasyim Asyari, which would give him great spiritual powers. After several days of prayer, the *kyai* led his students home disappointed because he had not found the *tasbih*. One student asked what the *kyai* had been looking for and was told, for the first time, that the *kyai* sought the *tasbih* of Hasyim Asyari. The student related how while he was praying and doing *zikir* (chant, lit. remembrance of God), a man came to him and showed him where a *tasbih* was hidden, but because he did not know the *kyai* was looking for it, he did not get it or mention the incident.

Because it encompasses secular education, traditional *pesantren* education, and a pilgrimage site, Tebu Ireng has been for many Classicalist Muslims in Indonesia a model for how to engage modernity and the state. However, it is a contested model, both within the Tebu Ireng community as well as in the wider *pesantren* community. The nature of the disagreements shed significant light on the processes in which both modernity and tradition are being (re)invented. As Tebu Ireng has interacted with processes of modernization and globalization, the grounds have been made into a sacred space that is meant to serve as an *imago mundi*, or a microcosmic model for how the world should be (cf. Eliade 1958:373).

Tebu Ireng enjoys a special place in the *pesantren* milieu. It is referred to as the *kiblat* of the *pesantren* world. *Kiblat* literally means the direction of prayer, of which there in only one: toward the *kabah* in Mecca. Used metaphorically, the term *kiblat* suggests a number of possibilities: a pilgrimage site, an *axis mundi*, and a focal point. I argue here that in modern Indonesia, Tebu Ireng, a *pesantren* established by a famous Sufi *Syehk* (master) and *fikih* (Islamic law) scholar,

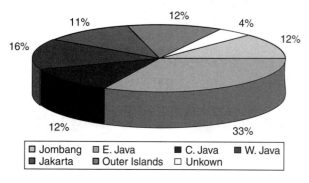

Figure 2.1   Parental residence of Tebu Ireng *Tsanawiyah* graduates 1995

is a popular pilgrimage site and in its layout provides a model for the ideal relationship between Islam, the nation, and modernity.

Tebu Ireng is clearly Indonesia's most famous *pesantren*. Founded as a "seminary" in 1899, it now provides a range of secular and religious education from the junior high through the college level. The founder of Tebu Ireng, Hasyim Asyari is regarded as a saint by Classicalist Muslims in Indonesia. He was also one of the cofounders of *Nahdlatul Ulama* (NU; Revival of the Scholars). His descendants continue to play an important role in both NU and the nation. His grandson Abdurrahman Wahid (Gus Dur), the former President of Indonesia and former general chairman of NU, have played and continue to play important roles in the Indonesian public sphere.

Tebu Ireng has about 1,500 students, from all over Indonesia (see figure 2.1). The students are all male, but Tebu Ireng is part of a complex of family *pesantren* that includes *pesantren* for female students. Tebu Ireng gives a slight emphasis to secular education over traditional *pesantren* education. Tebu Ireng has a rich history that is intertwined with that of the Republic of Indonesia. This national range of influence is reflected in the fact that Tebu Ireng's students come from all over Indonesia. Further, the grave of its founder, in the heart of the school is an important pilgrimage site that draws several thousand visitors each month.

History of Tebu Ireng

Tebu Ireng was founded in 1899 by Hasyim Asyari, who came from an *ulama* family. He is said to have possessed unusual intelligence and by the age of 12 was teaching in his father's *pesantren* (Dhofier 1995:2). According to legend, after serving Kyai Kholil of Bangkalen, Madura for several years, he had his master's considerable knowledge mystically transferred to him by a blow to the head. Following this episode, Hasyim Asyari studied in Mecca for eight years under some of the most notable Islamic figures of the nineteenth century (Dhofier 1995:3, 8; Rachman 1997:215).

Tebu Ireng was founded in response to negative effects of the Cukir sugar factory on the morals of the people; the area around the sugar factory was well known for drunkenness, prostitution, and gambling. After a few years of trying to run a *pesantren* in such an environment, he realized that before a *pesantren* could succeed, the trust of the community had to be won. This often involved giving people economic alternatives to working in the sugar factory, through job training and granting start-up capital (i.e., a horse and buggy to run transport services). Out of gratitude, many people accepted Hasyim Asyari's invitation to join in prayers and religious instruction. From the beginning, Tebu Ireng was involved in trying to guide Indonesia's economic development along Islamic guidelines. Today, Tebu Ireng's contributions in this vein are largely limited to education, although, for a time, Yusuf Hasyim served as the director of P3M, an NGO dedicated to community development through *pesantren*.

In the anticolonial movement, Hasyim Asyari, his students, and family played several roles. First, Hasyim Asyari was a cofounder of NU, which had an anticolonial orientation. Second, Hasyim Asyari wrote *fatwa* (considered legal opinion) that stated that all things Dutch, including dress must be rejected as belonging to unbelievers (*kafir*). He advocated a style of dress and lifestyle that would symbolically separate the Indonesians from the Dutch. Third, many Tebu Ireng *santri* were members of the *Hizbullah*, the Islamic anticolonial forces, and Yusuf Hasyim was a lieutenant. Finally, at a point when all hope was lost, General Sudirman, the commander of the Indonesian forces, asked Hasyim Asyari to declare that the war against the Dutch was jihad. He gave a *fatwa* doing so and requiring all Muslims within 80 kilometers of the fighting to participate (Arifin 1993:81; Syihab 1994:30). This declaration restored the hope and confidence (*semangat*) of the Indonesian people. For his actions, Hasyim Asyari is recognized as a national hero (*Pahalawan Pegerakan Nasional*).

Two identity claims are made in the above versions of Tebu Ireng's history. First is the importance of Tebu Ireng, and *pesantren* in general, in the establishment of Indonesia. The story of Hasyim Asyari's jihad *fatwa* claims that without *pesantren*, Indonesia could not have won its fight against the Dutch. This point is reemphasized in the recounting of the role Tebu Ireng *santri* and the Hasyim Asyari family played in the Hizbullah and the Indonesian National Army during the war for independence. The second claim is that Tebu Ireng, and *pesantren* in general, should continue to be part of Indonesia's national development, and that only *pesantren* can ensure that such development will be balanced by traditional values and morality. These claims are important in the discourse about Indonesian identity; they state that the best way to be Indonesian is to be a good Muslim.

Tebu Ireng: Pilgrimage Site and Imago Mundi

The vast majority of visitors to Tebu Ireng come to make pilgrimage (*ziarah*) to the graveyard. The Tebu Ireng visitor's office keeps a guest book in which visitors list their reasons for coming. In a random sample of months, over 80 percent of the people accounted for in the guest book came for *ziarah*. An average of 1,000 people came for *ziarah* in each of the sample months. One local observer feels that this number is far too low because not all pilgrims register in the visitor's office. He estimates upwards of 3,000 people come each month and that as many as 10,000 came during *Muharram* (the first month of the Islamic ritual year) 1416 A.H. (May 1995 C.E.).

As an institution that has undergone a transformation from being a traditional center of Islamic learning, famous for training *ulama* to a religious boarding school primarily for junior high and high school students, Tebu Ireng's landscape reflects a tension between maintaining tradition and modernization.

Tebu Ireng occupies about 20 hectares (about 200,000 square meters) of land in three separate blocks; although in absolute terms this is a small landholding, it is quite large on densely packed Java. The first block is the main *pesantren* complex, which occupies 2.4 hectares. The second block, where the *madrasah*, the schools and the sports field are located, is 8.6 hectares. The third block is 9 hectares of wet-rice agricultural land (*sawah*) located about nine kilometers from the main complex (Dhofier 1980b:150). The first and third blocks were dedicated as *wakaf* (religious property) by Hasyim Asyari. The second block was bought over time by the *pesantren* starting in 1974. The main complex is surrounded by a 2.5-meter high wall with main entry gates on the east by the main road, and in the west leading to the schools (see figure 2.2).

Because of Tebu Ireng's special place in the Indonesian Islamic community, all of Tebu Ireng is seen as sacred. Slightly more than 11 hectares of the total 20 hectares have been dedicated as *wakaf*. The rest is owned by the *pesantren* rather than by individuals. In short, all the land is committed to meeting the religious goals of the institution.

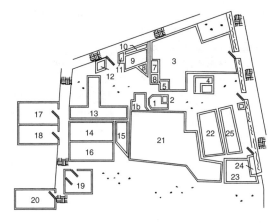

1. Mosque
1b. Graveyard
2. Student Affairs Office
3. The *kyai*'s house
4. Small store and *Bank Rakyat Indonesia*
5. Overnight guest room
6. Security office
7. Staff dining room
8. Stage (with TV)
9. MAK dormitory
10. General bathing room
11. The former house of Kyai Wahid
12. Skills center
13. *Tsanawiyah* building
14. Cafeteria
15. Courtyard
16. Girls' *pondok*
17. Junior high school
18. Senior high school
19. *Aliyah* building
20. IKAHA
21. Dormitory
22. Volleyball court
23. Classrooms
24. *NU-SUMMA* bank
25. *Pesantren* office and library

Figure 2.2   Grounds of Tebu Ireng

Clearly, however, some of the space (bank, kitchen, library, store) is profane. It is the inclusion of profane spaces into the sacred geography that provides a model for Muslim living. The message is clear: modern conveniences are acceptable as long they are kept in their proper place.

The Mosque–Graveyard Complex: The Sacred Center

Mircea Eliade, citing Levy-Bruhl, suggests that a space may be sacred because a mythical hero did something or built something there. Further the hero may even be buried there (1958:367). Hasyim Asyari is recognized as both a national hero and as a Sufi saint. He and his sons built a school of international repute. He and a number of family members are buried inside *pesantren* grounds. The mosque and the graveyard at the center of Tebu Ireng form the most sacred space. As a model for the life of individuals, the *umat* (Islamic community), and the nation, the placement of the mosque–graveyard complex suggests the importance that should be placed on two dimensions of Islam: normative piety as exemplified by the mosque, and mysticism as exemplified by the graveyard and the activities that take place there. Interestingly, the village of Cukir was known for its debauchery prior to the establishment of Tebu Ireng. Now Cukir, and the greater Jombang area in general, is known as a center of Islamic learning and pilgrimage. The landscape of Tebu Ireng proudly proclaims that both normative piety and mysticism must be part of modern life and Indonesian statehood, and it demands that they be placed firmly at the center.

The Mosque

In Islam, a mosque defines a community because it is the center of prayer, and praying together defines a community. The nature and shape of the mosque says something about the community. The mosque is of the distinctive three-tiered pagoda roof style found in Indonesia (Java and Lombok) and Kerala (Woodward 1989:55–56). Because Sunan Kalijaga, the most famous of the Wali Songo (the nine saints who brought Islam to Java) is said to have used this style of roof in building the Demak mosque, this style is called the Wali Songo style. The three tiers are said to symbolically represent *shariah, Tariqah,* and *Hakikat* (Law, Mysticism, and Truth).

Like all mosques, this one is oriented toward the *kabah* in Mecca. The *kabah* contains a black rock believed to have fallen from the throne of God directly above its current position. It is hence the Islamic *axis mundi,* or center of the world. Concerning the direction of prayer (*kiblat*), Annamarie Schimmel writes,

> The one direction of prayer around which the people of the world are place, as in were, in concentric circles has been and still is the most visible sign of the unity of the Muslims; it is, so to speak, the spacialization [*sic*] of their belief in one, and only one, God. (1991:164)

Although all Muslims pray toward Mecca, in what direction Mecca lays is subject to some interpretation. In the Babad Tanah Jawah, Sunan Kalijaga is said to have set the *kiblat* by putting one hand on the central pillar that supports the roof and one hand on the *kabah* in Mecca (personal communication, Woodward, March 1999). Hence, Classicalists pray due west because this is what the *kiblat* was determined to be in Indonesia prior to the advent of magnetic compasses. This is the direction of the *kiblat* of the Tebu Ireng mosque.

At the entrance to the open-air verandah of this mosque, stands a large drum (*beduk*) used to call the faithful to prayer. Second, local lore holds that when Hasyim Asyari built this mosque he said that as long as it stood, *santri* would find knowledge and *barakah* (blessing/power) at Tebu Ireng. The verandah of the mosque has been extended twice, but the original structure still stands. Because of the special nature of the mosque, some people have been antagonistic toward the building of additional mosques for the *pesantren*'s use.

However, maintaining the original mosque has come at a high price. A maximum of 1,000 people can fit under its roof (including extensions), therefore it is not possible for all the 1,500 *santri* to pray together. Several *pesantren* officials acknowledged that this means that *salat jamaah* (communal prayers) are, in practice, no longer required. The lapse in this requirement at Tebu Ireng is seen by traditional (*salaf*) *kyai* as a decline in its spiritual and religious authority. Yusuf Hasyim said that he was planning to build an *aula* (meeting hall) next to the schools so that students can do *salat jamaah* without disrupting their school schedule. However, the Hasyim Asyari mosque still plays an important role in the local religious practice. For Friday prayers, the neighborhood men come to pray. Once the mosque has filled, the men use their prayer rugs and fill the yard, all the way back to the main gate. By praying together in this mosque, although crowded and uncomfortable, the participants are claiming membership in a community that is maintaining tradition.

Another important use of the mosque is teaching. In the evenings, the mosque is still used for *ngaji* (the teaching of Classical Islamic texts). During school breaks, Gus Isyom, the founder's grandson, teaches a few hundred students including *santri* as well as outsiders. These sessions start after *salat Isyak*, the last prayers of the day and continue for several hours, sometimes until midnight.

Normative piety is central to *pesantren* practice of Islam. Keeping to the daily prayers, studying the Qur'an, and keeping *shariah* are important elements in the practice of Islam advocated by most, if not all, *pesantren* people. However, this is not the end of Islam for *pesantren* people. Islam is much deeper and richer than normative practices alone and this depth and richness comes from Sufism. In the landscape of Tebu Ireng, mystical practice is mapped out in the form of the graveyard.

The Graveyard

To the west of the mosque is the graveyard of Hasyim Asyari and his family members. Sacred geography places this graveyard between the believers and Mecca when they pray. Because the dead are buried facing toward Mecca, the placement of the graveyard suggests that the great *kyai* buried there are still leading the community in prayer. Almost all of the *pesantren* I visited had the founder's grave as an important ritual site, whereas only Tebu Ireng placed it between the mosque and Mecca. This is, however, the pattern found at the tomb-mosque of at least two of the Wali Songo, namely Sunan Kudus and Sunan Ampel. Woodward reports that this pattern was also found at the *Mesjid Besar* (Grand Mosque) in Yogya before the graves were destroyed by Muhammadiyah people (personal communication, February 1997).

People visit graves to obtain *barakah*, which has existential qualities and thereby can linger in the body and is transmitted to the area around the tomb (Geertz 1968:49). The tombs of men (and sometimes women) possessing *barakah* form pilgrimage sites throughout Java and the rest of the Islamic world (Mottahedeh 1985:23; Okello Abungu 1994). In addition to the Prophet Muhammad (Schimmel 1991:167), this category includes famous *kyai*, the founders of Sufi orders, martyrs, and pious kings.

Grave visitation (*ziarah*) is an important part of popular piety in Java (and elsewhere in the Islamic world). Every Thursday night, thousands of pilgrims visit the royal graves at Imo Giri and Kota Gede in Central Java. In East Java, the graves of two of Diponegoro's[1] soldiers form a popular pilgrimage site at Gunung Kawi. *Tahlilan* is an important aspect of grave visitation in general. *Tahlilan* involves the recitation of the first half of the *Shahada* (confession of faith; "*la ilaaha illa 'illah*" (there is no god but God)). *Tahlilan* is doubly meritorious as both the person who performs it and the person for whom it is performed receive merit (*pahala*). At Tebu Ireng, *santri* perform *doa* (supplication) and *tahlilan* on Thursday nights. Thursday nights or the eves of Friday are special in Islam. In Javanese Islamic practice, at least, days are reckoned to be from sunset to sunset and so Thursday night begins the Sabbath day of Friday. Beyond Java, special observances are found in Iran (Mottahedeh 1985:138) and elsewhere.

As mentioned, one reason a saint's grave is visited is to absorb his *barakah*. Another explanation for making pilgrimage to graves is the efficacy of prayers made there. Several *kyai* and *ustadh* explained to me that the dead saint serves as an intermediary. The notion of the buried saint serving as intermediary is found elsewhere in the Islamic world, namely, Kenya (Okello Abungu 1994). They used the analogy of needing to take care of some official business and asking a friend or family member, who knew the official in charge, to approach him on one's behalf. The official is more likely to listen and take action if a friend, rather than a complete stranger, brings the matter to his attention. Likewise, Allah loves the saint and is more likely to hear and answer the prayers of those who draw near to him. Dale Eickelman reports the value of visiting a saint's grave in Morocco in almost the exact same way (Eickelman 1976:161–162). Similarly Pederson reports that in early Islam, tombs were seen as places where "prayer is heard" (1953:335). Another popular practice at the Tebu Ireng graves involves copies of the Qur'an. People will take a copy that has been at the graves, absorbing *barakah*, and leave a replacement copy.

The mosque–graveyard complex forms the physical and symbolic center of the *pesantren*. This complex is, then, representative of traditional Javanese Islamic piety. It reflects three dimensions of piety: *shariah*, scholarship, and mysticism. It clearly demonstrates that these three dimensions must be balanced, that one cannot exist without another. Other models of Javanese Islamic piety suggest that it is possible to transcend *shariah* by mystical practice and therefore no longer be required to uphold it (Woodward 1989:107). The leadership at Tebu Ireng, and indeed most, if not all, *pesantren* people would reject such a model in favor of the one modeled in the mosque–graveyard complex at Tebu Ireng.

Secular, Modern Space

Still within the sacred space of Tebu Ireng is what first appears to be profane space. By incorporating profane spaces within the sacred complex, the landscape of Tebu Ireng is modeling what the relationship should be between Islam and modernity. At the center of life should be Islam, and as already discussed, there should be a balance between *shariah* and mysticism. With Islam as central, Muslims may engage in all aspects of modernity including education, medicine, banking, marketing, and so on. But since the space in which these activities are conducted are on the grounds of Tebu Ireng, they are within sacred space. So, these profane activities and places are transformed by their relationship to a sacred center.

Student Housing

Surrounding the mosque–graveyard complex on three sides are student dormitories. In the dorms, students, sleep, study for school, receive some *ngaji* lessons, and study Arabic texts. Interior and exterior walls are coated with heavily diluted paint, which allows the accumulated effects of years of heat and humidity to show through. Rooms are small and crowded. Students do not have permanent beds, but sleep where they can, sometimes first rolling out a mattress, but often sleeping directly on the floor. The carpets, if any, are thin and worn.

Each student is given a small square locker to store a change or two of clothes, their toiletries, and school supplies. Across the balconies of many of the dorms are clotheslines drying garments washed by the students. The floors of the dorms are considered pure (*suci*) and no shoes are allowed to be worn on them. In fact, small stepping-stones of pure flooring material lead between some of the dorms and to the common bathing facilities shared by all students. Even by village standards, these are simple living conditions. This simplicity is by design; it promotes an ascetic lifestyle that is a traditional part of *pesantren* education.

Library and Office Space

Southeast of the mosque are the library and the *pesantren* office. The current library was built in 1974 and holds several thousand volumes, many donated by The Asia Foundation. They are in English and largely unread. The first library at Tebu Ireng was established in 1934 by Wahid Hasyim and had about one thousand volumes, mostly *kitab kuning* (lit., yellow books; Classical Islamic texts). However, in order to expose students to the outside world, this library subscribed to many magazines and newspapers. To this day, the library at Tebu Ireng subscribes to a wide range of magazines and newspapers, which the students are encouraged to read to develop a broad worldview. This is in sharp contrast with more traditional *pesantren* that forbid any outside reading material. The library is well used and includes an historical archive that holds the *kitab* and journals of Hasyim Asyari in Arabic as well as copies of every research project conducted on Tebu Ireng, now including my own dissertation. Another reflection of this greater openness to outside material includes posted newspapers and a television in an open-air courtyard, which is shown on Thursday nights after *tahlilan* at the graves.

Banks and Other Accommodations to Modernity

Just inside the gate is a branch office of *Bank Rakyat Indonesia* (Indonesian People's Bank), which facilitates the transfer of funds from distant parents to students for their allowances and their *pesantren* expenses. Also in the main complex is a branch of *Bank Perkreditan Rakyat NU-SUMMA*, where students and faculty at Tebu Ireng, as well as many neighbors, have savings accounts. This bank also provides loans to small entrepreneurs (Arifin 1993:59). *Bank Perkreditan Rakyat NU-SUMMA* is an interest-charging banking enterprise based on cooperative efforts between NU and the SUMMA banking group. The presence of an interest-charging bank inside a *pesantren* is a controversial move toward what some consider the dangerous side of modernity; the Qur'an has a strong prohibition on *riba* (interest, usury):

> Those who swallow down usury cannot arise except as one who Shaitan has prostrated by (his) touch does rise. That is because they say, trading is only like usury; and Allah has allowed trading and forbidden usury. To whomsoever then the admonition has come from his Lord, then he desists, he shall have what has already passed, and his affair is in the hands of Allah; and whoever returns (to it)—these are the inmates of the fire; they shall abide in it. (Qu'ran 2:275)

Hefner states that the NU leadership has attempted to define *riba* as only those forms of interest that have exploitative rates. However, they have not succeeded in allaying concerns about *riba* amongst NU's rank and file (1996:296). The acceptance of interest-based banking within a *pesantren* is a clear statement about how the Indonesian Islamic community should adapt to capitalism. It is, however, a statement that few *ulama* can accept.

Roughly west of the mosque–graveyard complex is the general kitchen. By May 1995, all students ate here, a change that reflects aspects of modernity. In traditional *pesantren*, students cooked for themselves or in cooperative groups; this was seen as part of the training in *kemandirian* (self-sufficiency). Because general education has reduced the amount of time for religious education, Tebu Ireng, and many other *pesantren*, have sought to regain it by preparing food for the students. However, since students are no longer cooking for themselves, this part of the traditional training in *kemandirian* has been lost.

Northwest of the graveyard is a school compound that now houses the *Tsanawiyah* (religious junior high). This compound has a gate that separates it from the main complex. Also in this complex is a house that is referred to as Abdurrahman Wahid's house, although it is occupied by a Tebu Ireng functionary. Wahid, who is the grandson of Hasyim Asyari, is the past general chairman of the NU, the largest Islamic organization in Indonesia and the past president of Indonesia. He is well known and well respected. For this reason people pointed out to me that this house was *his*. Later in this chapter, I will show how this purported ownership became important in a dispute over the public use of sacred space. Adjacent to the house is the north gate.

Across the road from the north gate was a new *losmen* (small simple inn); the building and the land are owned by Yusuf Hasyim's daughter.

However, this establishment is staffed by Tebu Ireng *santri*. This five-room inn is designed for those visitors to the *pondok* who want a more quiet and comfortable environment than the crowded free guest rooms in the *pesantren*. This inn reflects a change in *pesantren* clientele. Many who stay there are fairly wealthy parents from Jakarta, Surabaya, and other urban centers. It also reflects a trend in which the family of *kyai* will run businesses catering to the needs of *santri* and their parents. However, more traditional *kyai* criticize this practice, arguing that those who keep it are more concerned with profit than with religion. These detractors argue that any such support services should be run by the *pesantren*'s neighbors and in this way the *pesantren* can contribute to the economic development of the village.

Schools

To the west of the main complex are the compounds for the secular junior high and high schools, called *Sekolah Mengenah Pertama* (SMP; First Middle School) and *Sekolah Mengenah Atas* (SMA; Upper Middle School) respectively. Outside the high school, a large sign proudly boasts that the school has biology, mathematics, and computer laboratories. Further west is the *Madrasah Aliyah* compound and the campus of Institut KeAgaman Hasyim Asyari (IKAHA), which was originally established as Universitas Hasyim Asyari (UNHASY) in the 1970s.

According to Pak Mushin, a senior official at Tebu Ireng, Hasyim Asyari foresaw the need for Tebu Ireng's *santri* to play a role in the growth and development of Indonesia. In this regard, Indonesia would need not only *ulama*, but also people with a wide range of skills, albeit with a good moral foundation. Hence, the point of Tebu Ireng's curriculum now is to create people who can help the *umma*, who have a moral and religious basis, so that Islam can thereby grow. Pak Mushin concluded that because Hasyim Asyari saw the need for Islamic education to fit the needs of the time, he would probably approve of the changes in Tebu Ireng.

Wahid Hasyim, Hasyim Asyari's eldest son advocated the introduction of even more general subjects. He argued that the majority of students in *pesantren* did not want to be *ulama*, so it was a waste of time to teach them the Classical texts. They should instead be given a general education with a religious foundation (Dhofier 1980b: 159–160).

Wahid Hasyim wanted *pesantren* education to use modern educational techniques including tests, class levels, and evaluations. Therefore, in 1916, the *madrasah* pedagogy of formal class levels, long popular in other parts of the Islamic world, was introduced to Tebu Ireng and to Java. He further argued that without knowledge of secular subjects, Islam would be defeated. Therefore, in 1919, the *madrasah* added mathematics, Malay, and geography (Dhofier 1980b:157). In 1926, other secular subjects, such as history and Dutch were added. It should be noted that these changes were concurrent with emergence of non-*pesantren* schools sponsored by *Taman Siswa* and *Muhammadiyah*.

Hasyim Asyari initially opposed his son's proposed changes and believed "that such a radical reform would create uneasiness among *pesantren* leaders" (Dhofier 1980b:160). In spite of Hasyim Asyari's disagreement with Wahid Hasyim's ideas, he allowed his son to establish the *Madrasah Nizamiyah*, with a curriculum of 70 percent secular subjects in 1934. In the 1950s, Wahid Hasyim reorganized the Tebu Ireng *madrasah* into a pattern now used by the national *Madrasah* system: *Madrasah Ibtidaiyah* for elementary school children, *Madrasah Tsanawiyah* (Junior High School), and *Madrasah Aliyah* (High School). The Tebu Ireng *madrasah* in 1950s also had a *Madrasah Mu'allamin*, or a school for training high school religion teachers. This was paralleled by the *Sekolah Guru dan Hakim Agama Negeri* (Government Schools for Religious Teachers and Administrative Staff of Islamic Courts) and the *Pendidikan Guru Agama Negeri* (Government Islamic Teacher's Colleges, forerunner of IAIN) established by Wahid Hasyim in his role as the Minister of Religion (Dhofier 1980b:163–164). Dhofier argues that this reorganization brought the downfall of Tebu Ireng as an institution of advanced religious studies (1980b:164–165). In time, Tebu Ireng was no longer able to produce students who were competent in reading and discussing *kitab kuning*. Without being able to examine *fikih* text, the graduates of Tebu Ireng are unable to become Classicalist *ulama*.

Tebu Ireng's involvement in state projects is partially demonstrated by its educational evolution. Further, next to the schools is a helicopter landing area that is said to be used for visiting government dignitaries. The presence of such a facility, whether regularly used or not, makes a statement about the relationship between the Islamic community (*umat*) and the government.

To summarize how Tebu Ireng serves as an imago mundi, imagine the pilgrim's experience there. Pilgrims enter the *pesantren*, all of which is dedicated to religious purposes (*wakaf*), and immediately see a bank, a telecommunications office, and a library. They proceed past student housing to the mosque–graveyard complex and there perform a number of rituals. As they leave, they pass through the back gate and encounter government curricula schools and a helipad for government visitors. As they move through this landscape, they see a model for how a Muslim should live in the modern world. This model requires traditional piety and mysticism to be an important part of an Islamic practice that includes both religious and nonreligious scholarship and good relationships with the existing government, even if it is non-Muslim.

Use of Sacred Space in Family and National Politics

The use and restriction of uses of this sacred space was a part of the political maneuverings that anticipated the exit of Suharto. (In 1995, the expectation was that Suharto would die in office, but that it was just a matter of time before the Suharto era came to an end.) In 1995, Abdurrahman Wahid made a number of public appearances with Megawati Sukarnoputri, the daughter of Indonesia's first president. Megawati was also the head of the Indonesian Democratic Party who was removed amongst allegations of a government conspiracy against her. Indeed at the time, many said that a Megawati–Abdurrahman alliance could easily challenge the government.[2] The public appearances included *ziarah* to the graves of their parents. The journey to Blitar to perform *tahlilan* at the grave of Sukarno occurred with no difficulties.

By contrast, on the July 28, 1995, when Abdurrahman and his contingent attempted to make *ziarah* to the family graveyard they found the gates to Tebu Ireng locked. There were *santri* guarding the gates with a walkie-talkie in contact with one of the headmaster's sons. Abdurrahman and his party entered the *pesantren* complex through the house that everyone acknowledged as his, although it is occupied by someone else. They thought that the gate between the *Tsanawiyah* and the central grounds was locked and so they laid out mats and did *tahlilan* there in the *Tsanawiyah* grounds adjacent to the graveyard.

There were a number of interpretations of this event. A Tebu Ireng neighbor and one-time *santri* interpreted the event as a direct attempt by the headmaster, Yusuf Hasyim to keep Abdurrahman out of Tebu Ireng. He interpreted Hasyim's objection to the event as being that the *pesantren* would be used for political interests (*kepentingan politik*). Further, the major players (Megawati and Abdurrahman) had recently been censured (*tercekal*) by the government. Yusuf Hasyim, according to this interpretation, wanted to guard the partnership (*kemitraan*) between Tebu Ireng and the government. And indeed, Hasyim had recently come out in support of Suharto seeking reelection in 1998.

One Tebu Ireng teacher in attempting to counter rumors that Abdurrahman was deliberately kept from his mother's grave, suggested that the *santri* guarding the gate were supposed to communicate that

Abdurrahman and a small party would be allowed to enter through the family gate to perform a *haul* (memorial service held on or near the anniversary of the person's death) and *tahlilan* for his mother. However, the large following (upward to 1,000) would have to stay outside. Apparently this message was not communicated and so Abdurrahman found another means to his ends. This teacher further argues that the gate between the *Tsanawiyah* complex and the grave-yard was not locked but only appeared to be. I did not see the organizers at the gate or whether they tried to open it. It is possible that the gate was not locked and that the event organizers only thought it was locked, as the teacher claims. I have on occasion walked up to this gate and thought it was locked, when in fact, the chain was merely wrapped around the gate and not actually locked. This interpretation suggests that the lockout was merely a misunderstanding and not indicative of any internecine struggle. Of course, this needs to be put in the context that after Abdurrahman won the election at NU's 1994 Muktamar (National Conference), his uncle Yusuf Hasyim lead the unsuccessful initiative to have the election declared improper (*tidak sah*) and have Abdurrahman removed from office.

When I asked Yusuf Hasyim about the event, he described it as a miscarriage of protocol. He blamed the organizing committee of the local chapter of PMII (NU College Student Organization) for not obtaining the proper permissions, namely his. He showed me three documents: the invitation for the Haul, the invitation for a *Dialog Kebudayaan* (cultural discussion) at Tambak Beras (another Jombang area *pesantren*), and the letter he sent to the local police. He pointed out that the invitation was addressed to him, but not signed by him although it was taking place in Tebu Ireng. He felt that as the official headmaster of Tebu Ireng, permission should have been sought from him.

Yet as he continued to describe the events and his response to them, the political interpretation that I had already heard from villagers started to emerge. He received the invitations at 8:30 A.M. and by 10 A.M. he sent a letter to the police because of what he described as his political instinct. He looked at the invitation for the *Dialog Kebudayaan* and saw that the same people signed the invitation for it as had signed the invitation for the grave visitation. Therefore, he felt that it was a political maneuver to connect Tebu Ireng with the event at Tambak Beras, which included a number of political figures. He said that he heard that a meeting with Megawati was not allowed to happen and that her PDI followers would come to Jombang for the two events including the one at Tebu Ireng. He said that someone needed to be responsible and so he wrote to the police.

I asked him if this was a political maneuver that was behind it. He did not give me a clear answer. However, he did say that if Megawati made appearances at both Tebu Ireng, and Tambak Beras, her position as an opposition leader would be stronger. He said that he wants nothing to do with any of the political parties because they are not democratic but rather decoration (*bukan demokrasi tapi dekorasi*). He continued that by appearing with Megawati, Abdurrahman was abandoning the *Kembali ke Khittah* 1926 (Return to the Spirit of 1926) movement that took NU out of party politics in the 1980s. For the rest of the Suharto era, NU was not involved directly in politics. Technically, this removal from politics remains although NU is associated with the political party *Partai Kebangkitan Bangsa* (National Awakening Party).

Sacred Centers in Post-Suharto Indonesia

So far, this material has shown that the landscape of Tebu Ireng spatializes a model, albeit contested, for how the Indonesian Muslim community should relate to modernity and the nation-state. However, it would be a mistake to compare this to Borobodor, Sukothai, or even the Yogyakarta Kraton. Tebu Ireng's physical layout was not designed to be a model of anything. It grew organically out of the evolving vision the Hasyim Asyari family had for the school and for Islam in Indonesia in general. Because Tebu Ireng grew out of an emerging vision for how Indonesian Islam in general, and *pesantren* in particular should engage the modern world, its layout spatializes that vision. In this way, Tebu Ireng can be said to be an accidental imago mundi.

Lest we conclude that this spatialized model for a modern Islam, this accidental imago mundi is in any way monolithic or even hegemonic, it is important to briefly highlight some of the disputed aspects of the model. As already mentioned, Tebu Ireng is considered by many to be the *kiblat* of the *pesantren* world. This metaphorical usage suggests that Tebu Ireng is, like Mecca, a pilgrimage site and that it is the focus of at least part of the *umat*. To be the *kiblat* of the *pesantren* world means that the educational developments in Tebu Ireng have been closely evaluated and sometimes copied by other *pesantren*.

Wahid Hasyim (Abdurrahman Wahid's father) advocated and initiated the teaching of nonreligious subjects in *pesantren*; Tebu Ireng was the first *pesantren* to do so. However, the degree to which it should continue to be the *pesantren* world's *kiblat* is subject to debate. Some hold it as an example of a good thing gone bad, in other words that Tebu Ireng has lost sight of the religious and moral foundations of *pesantren* education. Tebu Ireng is indeed a microcosm of the whole *pesantren* world; there are many competing voices, and each attempts to shape the discourse in a particular way. So too in the wider *pesantren* world, there are those who wish to maintain or return to more traditional ways, and those who are striving to be modern. The latter claim to balance tradition and modernity, but the former is unconvinced by their rhetoric. It is beyond the scope of the current effort to explore further the various dimensions of these debates.

In 1999, Tebu Ireng was one of three contestants for the sacred center of Indonesia in the political, economic, and spiritual crisis that is the immediate post-Suharto era. In part, political debates in 1999 were about which center the future Indonesia will be based on: the Westernized, and perhaps corrupt, Jakarta, the traditional sultanate of Yogyakarta, or Tebu Ireng, which manages to balance *shariah*, mysticism, and scholarship within its practice of Islam, as well as balance Islam with the needs of a modern state. And it is precisely the model of Tebu Ireng and *pesantren* in general that bolstered former President Abdurrahman's supporters. It is nearly impossible to separate Classical Islam in Java from *pesantren* and so the next chapter explores the nature of *pesantren* education and how modernity and tradition are negotiated within that system. In short, many *pesantren* have found it necessary to offer curricula that prepare their students for life and work in the modern world. This means necessary changes in the traditional components of a *pesantren* education.

Contested Purposes

The model for a modern Islam that is spatialized in Tebu Ireng is contested and is part of the general contested nature of the discourse surrounding Tebu Ireng. In this way, Tebu Ireng is very much a microcosm of the *pesantren* world. Tebu Ireng's vision for the future of Indonesia is far from unitary. This is in part due to how it has solved a common problem in *pesantren*: inheritance. Although religious property (*wakaf*) is not heritable technically, there is a tendency for *ulama* families to form dynasties. During the life of the founder, the *pesantren* is under his single control; it is said that a *kyai* is a little king. Upon the death of the founder, a situation emerges in which each son (or his children) and each daughter (or, more likely, her husband) feel they should have a say in the operation of the *pesantren* and there is no ultimate authority. An eldest son nominally inherits his father's role, but on the condition that he is found competent in religious matters.

In Tebu Ireng, Yusuf Hasyim, the eldest surviving son of Hasyim Asyari, is the director, even though it is generally acknowledged that he does not have the skills necessary to teach or debate religious texts. Yusuf Hasyim has compensated for this by forming a council of *kyai* to advise him on religious matters, and through his management of this council, he has come to prominence as a leader in the Islamic community (Dhofier 1980b:169–170). However, this creative solution has also strengthened the sentiments of other family members that they share in the right to guide the development of Tebu Ireng.

There are several competing visions for Tebu Ireng, the *pesantren* world, and Indonesia found in Tebu Ireng and the Hasyim Asyari family. In terms of education, some, like Yusuf Hasyim, wish Tebu Ireng to continue on its trajectory toward increasing secular education (although maintaining a religious base), although others, like Gus Isyom, wish to reestablish Tebu Ireng as center of *ulama* training. In terms of the vision for Indonesia, some like Yusuf Hasyim seem to have argued that the *pesantren* world should be closely affiliated with the Suharto regime, although others, like his nephew, Abdurrahman Wahid (eldest son of Hasyim Asyari's eldest son) favored a more distant relationship with that government.

Yusuf Hasyim argues that changing times, and with them changing needs, require *pesantren* education to change. At the 1995 graduation for the *Madrasah Aliyah* (Religious High School), he likened these changes to those found in public transportation. At one time, the Ford Rose was the dominant minibus, then Mitsubishi introduced the Colt that was cheaper to buy and had better gas mileage. The Mitsubishi Colt was replaced by a Suzuki minivan, which was cheaper still both in purchase price and in operation. If *pesantren* are not competitive they will be driven out of the market; parents will not send their children to schools that do not give them a chance at success. He said that Tebu Ireng is not limited to educating religious leaders and in the last 20 years, general education at Tebu Ireng has yielded many alumni who work in a variety of fields, including in many government offices.

Yusuf Hasyim's use of mercantilist logic to justify changes in *pesantren* indicates a departure from traditional *pesantren* argumentation. Although the use of analogy is a solid tradition, Yusuf Hasyim's argument relies completely on reason and market principles rather than on the citation of religious authorities. In contrast, other *kyai* who advocate general education and other modern innovations would justify them by citing an Arabic phrase that they translated for me as, "*melestarikan yang lama yang baik dan ambil yang baru yang lebih baik*" (keep the old values and practices that are good and take (only) the new ones that are better).

Yusuf Hasyim is cautious about how far changes should be allowed to go. In our discussions, he frequently recounted the story of Harvard

University as a cautionary tale. As he tells it, Harvard began as a religious institution, but has since lost its religious character. He argues that this must not be allowed to happen with *pesantren*; although *pesantren* must add new subjects and new forms of education, the fundamental religious character of *pesantren* must not be lost. He said that the secular school system is designed to make people smart (*pinter*), but *pesantren* want to create people who are both moral and smart (*benar dan pinter*). He said that the fate of Harvard will be prevented in *pesantren* by the presence of the mosque and the enforcement of the five daily prayers. However, he acknowledges that the current mosque is too small to allow this and wishes to build an *aula* (conference hall also used for prayers) to accommodate current needs. He said that from the *Madrasah Aliyah* (Religious High School) they are trying to produce *ulama* and *kyai*. But, from the SMA (National High School) they wish to produce scholars with a *pesantren* soul (*sarjana yang berjiwa dan semangat pesantren*).

However, not all are convinced. Dhofier states that because of the decline in religious and moral education at Tebu Ireng, it is no longer able to provide leadership to the Indonesian Islamic community. Dhofier's hope now lies with Tambak Beras and the descendants of Wahab Hasballah (personal communication, May 1995). When Dhofier studied Tebu Ireng, it had a dual paradoxical role as both a center of traditional Islam and a center of modernization. Dhofier notes that with the loss of Kyai Idris, the last Tebu Ireng *kyai* to train *ulama*, there was a "growing pessimism over whether Pesantren Tebu Ireng [would still be] capable of producing a new generation of qualified *kyai* who can develop new *pesantren*" (1980b:169). By the mid-1990s, there is little doubt left; Tebu Ireng is no longer a center of traditional Islamic scholarship and is instead mostly an agent of change bringing metropolitan culture to rural areas. As evidence of this, Imron Arifin offers the establishment of a bank within the *pesantren*, in spite of the fact that many *ulama* still see interest (*riba*) as something that is *haram* (forbidden) (1993:53). Arifin also cites the widespread use of pants, amongst both teachers and students, which violate modesty laws, instead of sarongs, as another indication of the change in the moral climate of Tebu Ireng.

There are many both within and around Tebu Ireng who would agree with Arifin's assessment that Tebu Ireng's role as an agent for modernization is no longer balanced by the role of being a center of traditional Islam (1993:54ff). One elderly neighbor, who is also a distant relative of Hasyim Asyari, stated that in times past, young men, who had yet to complete even their fourth year of study, would be sent out to serve remote villages as religious leaders. Even Yusuf Hasyim acknowledges that such proficiency in religious scholasticism is no longer found at Tebu Ireng, a fact he attributes to changing times and changing needs.

However, there are also those within Tebu Ireng who have noticed these problems and are working to correct them. Gus Isyom, a grandson of Hasyim Asyari, established a special program to train a handful of young men to become *kyai*. Some accuse him of being unrealistically nostalgic and wanting Tebu Ireng to return to the days of Hasyim Asyari. Gus Isyom asserts that because of globalization and technological development, it is neither possible, nor wise, for all of Tebu Ireng to do so. However, he argues that some parts of Tebu Ireng should strive to capture the spirit and moral values of those days. He also argues that in Hasyim Asyari's days, in addition to the role *pesantren* played in education, they also played a larger role as centers of intellectual development and social action, a role to which he wishes Tebu Ireng to return. He said that there is not much hope for this to come from the current educational system at Tebu Ireng. His hope is that Tebu Ireng will be able to strike a balance between its new role and its old one, with most of the *santri* in the general education programs and about 25 percent of the *santri* engaging in the type of educational system that existed under Hasyim Asyari.

Gus Isyom said he did not know all of what Indonesia needed for development or how to achieve it. He was quite certain, however, that Indonesia needs to develop true democracy; that is, equal opportunity in education, work, and in the political process. He said that *kyai* and *pesantren* are best suited to play a role in democratization because the very title of *kyai* is bestowed by the populace. Therefore, he argues, *kyai* must defend the disenfranchised masses. His uncle, Yusuf Hasyim agrees with this and has written,

> The bread of development has not yet been evenly distributed but has only been enjoyed by the upper crust of society. . . . As previously mentioned, social imbalance colours our social structure, especially in the uneven acquisition and distribution of natural and human resources . . . *Pesantrens*, having strong roots in the lower classes can be potentially liberating. (1987:11–12)

Gus Isyom laments that this role is not fully realized because many *kyai* have become elites through their relationships with power brokers (especially in the government). He said that *kyai* should return to their popular base; they should join in the peoples' activities, to give guidance and to be more accessible. Gus Isyom says that he does this with his *santri* by teaching regularly and participating in other student activities. Further, both in his teaching, and outside of it, he tries to address the student's problems in terms they comprehend. He wishes to restore the familial connection between *santri* and *kyai*, which he feels has been lost, at least at Tebu Ireng: *kyai* should be a father figure, not an administrator. Isyom argues that a close relationship between a *kyai* and his students is crucial in creating a moral commitment to the poor in the *santri*.

In Tebu Ireng there are two competing visions. One is to move Tebu Ireng, the *pesantren* world, the Indonesian Islamic community, and Indonesia into the twenty-first century. The concern in this vision is for economic success, "modernity," and development. It is modeled best by Yusuf Hasyim, who encourages his students to study science and technology and who uses a nontraditional style of argumentation. Although this vision is concerned with morality, it was difficult to get those affiliated with it to define morality, much less describe how to teach it. Following this vision has meant a change in Tebu Ireng's role. Once it was a school of higher Islamic education, on par with any *madrasa* in the Islamic world, and in some cases better. Now it can no longer play this role. The other vision for Tebu Ireng seeks to balance the push for modernity with a return to Tebu Ireng's role as a "graduate program" in theology and religious practice. This vision is modeled by Gus Isyom, who trained at Lirboyo, a *pesantren* considered by many to be the best contemporary *salaf pesantren*. Many in and around Tebu Ireng see Gus Isyom as the first member of the Hasyim Asyari family in a long time who has the caliber of religious prowess and knowledge to be a true successor to Hasyim Asyari. Abdurrahman Wahid is said to be of the same caliber, but his activities in NU have taken him away from Tebu Ireng.

Tebu Ireng is a model for the *pesantren* community. It is a contested model and that the contested discourse within Tebu Ireng is a microcosm of the *pesantren* community as a whole. The chief concern is how to engage education while maintaining the educational distinctiveness of the *pesantren*. Clearly, adding government recognized curricula reduces the hours available for traditional *pesantren* education. Many *kyai* are concerned with how exactly to balance modern secular education and religious training. Some have called for a slight change in focus from training religious leaders to training pious laymen. The next chapter turns to the debates within the *pesantren* community about the nature and future of *pesantren*.

Chapter 3

"Politics" by Other Means: Using Education to Negotiate Change

In the summer of 1992, I was attending a language program in Malang, East Java. During a weeklong break, one of my language teachers took me to a number of *pesantren* in East Java. I had, of course, already formed an interest in them based on readings, but had yet to see them "on the ground," so to speak. I have to confess that I was left more confused than anything else by that initial foray. In part, it was due to ethnocentric understandings of the terms *modern* and *traditional* that I held at the tender age of 26. Mas Yanto had taken me to a number of *pesantren* and told me briefly what he thought of them. We visited one that had huge gleaming buildings with ceramic tiles, apartment block buildings for the students dormitories, pedestal toilets, and even air-conditioning and a bed in the guest room in which we stayed. Yanto, who had never spent anytime in a *pesantren*, but was interested in attending one after he completed his college degree, told me that this *pesantren* was a very traditional one. He took me to another that was by comparison, rusticated. Although it had been nicely landscaped, some of the buildings were built of bamboo, many of them were simple concrete covered with a watered down paint that barely hid the grime and mold that accumulates in a tropical climate. Yanto told me confidently that this was a modern *pesantren*. When I finally inquired as to the criteria he was using, I found out that he was referring to the curriculum. The first focused only on religion, in fact only on the Qur'an and hadith while the second had added government curricular schools. It was then I realized that *pesantren* education was undergoing change. I also quickly discovered that not everyone would agree with Yanto's typology. Indeed, I came to discover that there was considerable debate about what properly constituted a *pesantren*, how best to categorize them, and what their future would be. When I returned to conduct my fieldwork in 1994, the question of educational change and identity became a major focus.

Central to the identity of the *pesantren* world is its history. This is not a particularly surprising or illuminating fact in itself, however, the contents of this history, or at least the parts that enter into contemporary discourse, make claims about the nature of *pesantren*. History is not merely an accounting of events, but an interpretive affair. Hence, the various claims explored here strive to show that *pesantren* are simultaneously Javanese, Indonesian, and Islamic.

In the *pesantren* world, it is generally agreed that the *pesantren* is a local institution that teaches universal Islamic practices and beliefs. However, exactly how the *pesantren* is local is subject to some debate; this debate invariably invokes history. The *pesantren* world in Java is nearly as old as Islam in Java itself. Both in print, and in oral tradition, *pesantren* are closely tied to the Wali Songo (the nine saints who brought Islam to Java). The first, if not the most famous, of the Wali Songo, Sunan Maulana Malik Ibrahim is said to have established the first *pesantren* in Java in 1399 C.E. in order to train *muballigh* (preachers) to further spread Islam in Java (IAIN Sunan Ampel 1992:21).

Abdul Gani, an *ustadh* at An Nur in Bululawang, told me that while Malik Ibrahim is known specifically for his use of *pesantren* to spread Islam and that each of the Wali Songo are known for other specific methods of proselytism, each established *pesantren* upon retirement from their other endeavors. The claim that all the Wali Songo had *pesantren*, which is supported by a few published accounts (IAIN Sunan Ampel 1992:22; Yunus 1979:217), serves to establish the contemporary *pesantren* world as inheritors of the Wali Songo.

No *pesantren* claim to have been founded by one of the Wali Songo, but all *kyai* are seen as inheritors of the role of the Wali Songo. However, there are some religious figures like Syehk Abdurrahman (discussed in chapter 1), who claim to be descendants of one of the Wali Songo. *Kyai* who affiliate themselves with men like Syehk Abdurrahman (e.g., Hasyim Muzadi of Al-Hikam) are drawing connections between themselves and the archetypes of *kyai*, the Wali Songo. Further, the *pesantren* world in general is seen as carrying out the role of the Wali Songo.

In Indonesia, the missionaries responsible for the initial Islamization of the archipelago are regarded as saints and are referred to as the Wali Songo, or the Nine Friends of God. The Wali Songo have great mystical feats attributed to them. Sunan Kalijaga is said to have meditated without ceasing for 40 years.[1] Another of the Wali Songo is said to have been able to fly to Mecca for Friday prayers and be back home in Java for his afternoon repast.

Several of the Wali Songo are known for making accommodations with local culture. Specifically, they are known for their use of *wayang* (shadow puppet theatre). One way this was done was by reinterpreting the Hindu epics. For example, in the Mahabarata, Arjuna has a secret weapon called the *Kalimasada*, which Javanese Muslims say is short for *Kalimah Shahada*, or the Islamic Confession of Faith. Some have argued that the highly stylized human forms in the Javanese *wayang* puppets reflect the influence of Islam that discourages the artistic representation of the human form. The Wali Songo are also known for incorporating the *gamelan* (percussion orchestra), the slit gong, and the *beduk* (large drum) into the call for prayer. The purported logic was that by using sounds that people already associate with large gatherings, people would be more interested in attending the prayers.

An important episode out of the Wali Songo legends is the story of Syehk Siti Jenar. Syehk Siti Jinar was one of the nine but he taught the heretical notion that since God is one, there was no separation between himself and God, and that he was therefore God. He was summarily executed; not for being incorrect in his assertion, but for teaching something that would confuse the common believer. This tale emphasized the importance of balancing the normative and mystical dimensions of Islam. It is also important to note that this is a retelling of the Persian legend of Al-Hallaj (Woodward 1989:102–103).

Another key Wali Songo is Sunan Kudus, who is said to have originated the *wayang golek*, the three-dimensional wooden puppet show, and for building the Kudus mosque. The cultural accommodation of the Wali Songo is further celebrated in a story that relates that because the residents of Kudus were predominately Hindu, Sunan Kudus built his mosque to resemble a Hindu temple and forbade his followers from eating beef. To this day, the Kudus area, now predominately Muslim, is known for the fact that the inhabitants do not eat beef but favor water buffalo.

The Wali Songo, Kalijaga in particular, taught Islam through local art and culture, specifically the *slametan* (ritual meal) and the *wayang*, or shadow puppet theater (Woodward 1989:96). To this day there are *pesantren* that sponsor *wayang*, *gamelan* orchestras, and other cultural events (Pranowo 1991). Pesantren Lirboyo, perhaps the most famous traditional *pesantren*, hosts an annual exhibition of Javanese martial arts (*pencak silat*) that draws crowds in the thousands.

Some claim that one of the local institutions adopted by the Wali Songo were Javanese Hindu–Buddhist monastery schools, which became the Javanese *pesantren* (Abdullah 1987:80; Jones 1991:19–20). Mustahin at Tebu Ireng strongly disagrees with those who assert that *pesantren* are derivative of Hindu religious schools and asserts that *pesantren*-style education began in the time of the Prophet.

Indeed, as institutions of Islamic learning, *pesantren* were and are by no means unique to Indonesia; similar schools are found throughout the Islamic world including Iran (Mottahedeh 1985) and East Africa (Reichmuth 1993), and have a deep and colorful history (Makdisi 1981). It is beyond the scope of the current work to give a complete history of Islamic education, but it is useful to highlight some aspects of it. Almost immediately, the Islamic community taught its members the skills necessary to read, memorize, and recite the Qur'an. Shortly afterwards came schools that taught hadith and the skills for determining which were sound and which were not. Eventually all but four of some five hundred of these schools disappeared. The remaining four became the four *madhab* (schools of jurisprudence) and lost their academic nature (Makdisi 1981:2). Early forms of Islamic education took place in the mosque in part because there was nowhere else for them. In time, the mosque became established as the ideal institution of learning (Makdisi 1981:12). In Java, the first building of any *pesantren* is a *musholla* or mosque.

In medieval Islam, learning took place in mosques in *halqa* (learning circles). According to George Makdisi, the medieval *mesjid jami* (congregation mosque) was not only a place of worship on Fridays but was also a place of learning in which professors of various Islamic disciplines and related subjects, like Arabic language and literature, were taught in a *halqa*, or learning circle. Interested students could attend any *halqa* that interested them. However, *halqa* on law had the restriction that only members of the *madhab* whose law was being taught could attend. But since Muslims may change their *madhab* at any time, students could attend the law *halqa* of their choice as well (1981:18–19).

Over time this form of education evolved into the *madrasa*, which was first and foremost committed to the study of law and therefore became the "institution of learning par excellence." The other Islamic sciences were studied as ancillaries (Makdisi 1981:9). Makdisi argues that the mosque school continued to be used for the teaching of the various Islamic sciences, including that of law, according to the wishes of the founder. The delineation between *madrasa* and mosque school were not clearly delineated in all places. J. Pederson, working in Cairo, saw no distinction between the *madrasa* and the mosque school (Makdisi 1981:20). R.A. Kern says that the late-nineteenth- and early-twentieth-century *pesantren* was an institution of advanced theological study that concentrated on *fikih* (Islamic jurisprudence) (1953:461). Although many changes have taken place in many *pesantren*, Kern's assessment continues to be true for the most traditional *pesantren*.

Others claim that *pesantren* are unique because they combine two institutions, the *funduq*, a place to study and practice Islamic mysticism and the *pesantren*, a place of Islamic scholasticism. Abdurrahman Wahid, in a public lecture, argued that the Malaysian version was only a place of Islamic learning and did not encompass the teaching of Islamic mysticism; therefore, the Malaysian *pesantren* has disappeared under the pressures of a changing world. Because of the composition of the Javanese *pondok pesantren*,[2] Zamaksyari Dhofier argues that Islam in Java did *not* bifurcate the Islamic sciences and so did not have two types of Islamic teachers: scholars (*ulama*) and mystics (Sufis). Hence the Javanese use the unique title *kyai* to "indicate a Muslim scholar who is both a master in Islamic theology and jurisprudence and a Sufi master" (1980b:31–32).

However this combination of intellectualism and mysticism is by no means unique in Islam, at least in Classical Islam. Pederson argues that because learning and manifestation of piety are inseparable in Islam, the *madrasa* (place of Islamic learning) and the *ribat* (place of Islamic mysticism) were frequently combined (1953b:305). Makdisi argues that such combined institutions were common by the twelfth century C.E. (1981:10), well before the first *pesantren* was established in Java.

The above evidence suggests that the Javanese *pesantren*, prior to the mid-twentieth century belonged to the category of *madrasa* and hence can be compared cross-culturally to other Islamic *madrasa*.[3] However, although it is more likely that the form of *pesantren* is Islamic, we cannot dismiss out-of-hand the claims that the *pesantren* is based on older pre-Islamic Javanese institutions. Those who make this claim about the history of *pesantren* are claiming a particular identity for the *pesantren* world today; the *pesantren* might be an Islamic institution but it is also a local one.

Another set of historical claims about *pesantren* further defines the identity of the *pesantren* community as quintessentially Indonesian. These claims concern the role of *pesantren* people, especially *kyai* in colonial resistance, the Indonesian War for Independence, and in post-Independence politics. Many nineteenth-century rebellions were led by *kyai* (Kartodirdjo 1966). *Kyai* and their *santri* were very much involved in Indonesia's struggle for independence. In the written and oral histories of many *pesantren*, is an accounting of the role played by its personnel during the war. For example, Hasyim Asyari of Tebu Ireng is said to have rejuvenated the independence effort by declaring that the war against the Dutch was jihad and required (*wajib*) for every Muslim within an 80-kilometer radius of the enemy. Other *kyai* lead their *santri* as guerrilla cells. This nearly universal claim of having some role in the war effort is a claim of identity; namely, that the *pesantren* community fought hard to be Indonesian.

It is interesting to note however that several prominent *kyai*, some of whom had fought in the war, expressed to me with great dismay that the government seems to be omitting their role in the war from official history. Indeed the official history depicted in the dioramas housed in the base of the National Monument (*Monumen Nasional*) confirms this concern. In the dioramas, the only religious expression that is shown as a force for nation building is *Muhammadiyah*. The diorama depicting *pesantren* shows *santri* gathered around a *kyai* reading a religious text; it gives no indication of the role the *pesantren* world played in the struggle for independence. In 1995, there was a seminar held to discuss strategies for ensuring that the role the *pesantren* world played in the War of Independence is not forgotten; clearly this is of concern to them. While *pesantren* people are trying to claim not only that they are Indonesian, but that Indonesia would not exist without them,[4] the government is downplaying these claims.

Curriculum and Identity

Lawrence Levine, in *The Opening of the American Mind* (1996), argues that debates about curriculum in American universities are about both social goals (what America should be) and social reality (what America actually is). The shift from classical curricula (Greek and Latin grammar) to Great Books and the "Western Civ" curricula, was occasioned by the political necessities of the World Wars (1996:64). Likewise, the introduction of women's studies, ethnic studies, and multiculturalism reflects changes in American society as a whole (1996:28). The debates about these changes have been acrimonious and heated, with some decrying them as a complete break down of American education that threatens the soul of America (Levine 1996:4–6). The debates about curriculum change in *pesantren* have been no less acrimonious, even if the issues (whether or not to teach biology, e.g.) are nearly nonissues in the American mind.

Originally *pesantren* were exclusively for the advanced study of religious knowledge. However, they now often include secular education, mostly at the junior high school and high school levels but sometimes at the elementary school level as well. This section outlines the issues in this curriculum debate, the details of which will be explored elsewhere in this volume.

In the nineteenth and early twentieth century, *pesantren* education was typified by each institution and *kyai* specializing in one area of knowledge. Hasyim Asyari at Tebu Ireng was known for hadith, whereas Pesantren Jampes of Kediri was well known for its *kyai* who were experts on Sufism (Dhofier 1980b:11). If a student wanted to become an expert in hadith he would study under Hasyim Asyari; for Sufism he would go to Jampes. This form of education provided in-depth knowledge of a particular field of study. In order to achieve a wider range of knowledge, the *santri* would move from institution to institution studying different fields in great detail.

Dhofier describes the wandering scholar tradition in the person of Wahab Hasbullah, who replaced his father as the head of *pesantren* Tambak Beras, and was also one of the cofounders of *Nahalatul Ulama* (NU). During his wandering, he studied under many famous *kyai*, such as Kyai Kholil Bangkalen and Hadratus Syaihk Hasyim Asyari. Further, he was classmates with several other young men who also became great *kyai* (Dhofier 1980b:15–20). This tradition created *ulama* that had both a solid foundation of training as well as broad network of contacts, both horizontal (with classmates) and vertical (with teachers). Whereas Dhofier describes the tradition of a wandering scholar as still having some depth when he conducted his research in 1977–1978, by the time I conducted research in 1994–1995, it was largely in abeyance. Something that only vaguely resembles it still existed; during Ramadan, *santri* from other *pesantren*, university students, and other young people (mostly men) might attend special Ramadan lessons.

Contemporary *pesantren* often offer the whole range of religious knowledge, keeping several teachers on staff who together teach various subjects. At *pesantren* that have adopted government curricula schools, students get a broad exposure to classical teachings. However, because the students are also enrolled in general education, the time they have for religious subjects is reduced. Hence, the question arises where does one find the in-depth religious training needed to become an *ulama*.

The curriculum found in contemporary *pesantren* can be divided into four basic types: *ngaji* (religious education), *pengalaman* (experience), *sekolah* (general education), and *ketrampilan* (skills training) and *kursus* (courses). The most traditional *pesantren* have just *ngaji* and *pengalaman*. *Sekolah* is a more recent development. Although *sekolah* in *pesantren* has roots as far back as the early years of the twentieth century, it emerged in force in the late 1970s. The history of *kursus* and *ketrampilan* is more difficult to follow and may be related to older "apprenticeship" practices, but has emerged in force more recently.

Ngaji *and Religious Education*

Ngaji (sometimes also called *pengajian*) is religious education and is both a form of education and a type of instruction found in *pesantren*. One can study *ngaji*, that is, learn how to read texts; one can also *ngaji* a text, that is recite and learn the meaning of it. The first form of *ngaji* is simply the study of how to read and recite Arabic texts, first and primarily the Qur'an. Studying *ngaji* is not limited to *pesantren*; many children first learn to *ngaji* from their parents, or from a tutor in their home or neighborhood. Basic *ngaji* focuses on the proper pronunciation and delivery of Qur'anic passages and does not include reading comprehension. It is considered the minimal amount of religious education one should have. Parents who are lax in their observance of Islamic law, still insist that their children learn to *ngaji*.

In the wider Islamic world, *pesantren*-like institutions (*madrasa*) are for intermediate or advanced religious training, however there are also children's schools called *maktab* or *kuttab* that teach basic skills. Like basic *ngaji* in Java, the lessons might be held in a school or in the homes of the student (Pederson 1953b:300). Once a student has mastered the basic skills they might continue their studies in a *madrasa*. Roy Mottahedeh reports such a pattern in Iran,

> The *madreseh* was not expected to give its students basic literacy or an elementary knowledge of arithmetic. Either private tutors, such as Avicenna's teachers of Koran and literature and the vegetable seller who taught him "Indian calculation," or small, unendowed neighborhood Koran schools, which lived on the fees they collected, still provided the very beginnings of education. What the *madreseh* was expected to give was a basic education in Islamic religious law. At the same time, from the very beginning it was intended that the *madreseh* should teach the student the relation of law to its sources, especially to the Koran and the accounts of what Mohammed said and did. (1985:90)

Likewise in Java, *ngaji* in a *pesantren* (*madrasa*) is far more than the minimum. Many, if not most, *santri* already have the minimal skills when they enter; those who do not, receive remedial instruction.

Traditional education in *pesantren* was self-directed and self-paced. The student would select various *kitab* (classical texts) and study them under the tutelage of a headmaster (*kyai*). Individual study between the student and the *kyai*, called *sorogan*, requires much patience and diligence on the part of the students. The student brings a text to the *kyai* and reads it in front of him to be corrected (Dhofier 1980b:20). This method may be used twice in a *santri*'s career; at its beginning and at its end. It may be used at the beginning of their studies, if they need help in basic skills. In this case, the students usually do not work directly with the *kyai*, but with an *ustadh*, or a senior *santri*. *Sorogon* is also used with advanced students, and in this case usually involves the *kyai*.

Sometimes groups of students study the same book in a method called either *wetonan* or *bandongan*. The following is a composite description of *wetonan* for advanced students at An Nur. The students who participated were mostly older students, over 15, however some were younger that leads to the conclusion that there was no minimum age but there was a minimum skill level—the ability to read and write in Arabic script.

The *santri* sat cross-legged on the floor along the edge of the *musholla* (small mosque) leaving the center open. Note that although lower levels of instruction are conducted in other areas of this partic-ular *pesantren*, the highest level of *wetonan* is done in the mosque. The students had loose-leaf copies of a *kitab kuning* and pens. Some had fine-tipped ballpoint pens; most preferred nibs and inkwells. There was much talk and gossip. When the *kyai* approached they shushed themselves. The *kyai* sat down, took attendance, and began the lesson. Armed with a microphone, he read a few sentences in Arabic. He then stopped, and, in Javanese, gave the *makna* (formal meaning), which is more a translation than an explanation. The students write the meaning between the lines in their own copy of the *kitab* in fine Arabic script Javanese. This dictation method is seen by some as indication of a lack of comprehension of the texts and hence as an indication of the imperfection of Javanese Islam (Geertz 1960a:178). However, Pederson reports that this teaching method is quite common throughout the Islamic world and does not indicate a lack of under-standing (1953b:306). It is also only the beginning of the learning process. Students will review and repeat the lesson on their own or with a friend. Also, groups of students, particularly the advanced ones, are expected to gather and debate the meaning of the text.

The *kyai* was called away to attend to guests, so the students reviewed on their own. One by one *santri* were called forward to read passages from the text. Others checked the veracity of what they have written with their neighbors. However, in the absence of the *kyai* (and know-ing that he will probably not return) the atmosphere changed dra-matically: some *santri* followed along with the reading; some laid their heads on the floor to sleep; some stretched out their legs; most of the senior *santri* moved toward the door so that they would not be hot; and my note taking became a focus of attention with nearby *santri* paying more attention to the strange foreigner than the review.

The next day, those who were absent the day before were singled out and made to sit in the center. After the lesson, those who were in the center were asked by the *kyai* if they had prepared, and those who answered in the negative became the subject of ridicule at the hands of the *kyai*. Public ridicule is a common form of discipline in *pesantren*. I once observed a dose of this punishment that included ear pulling in addition to the ridicule. To make the humiliation greater, the *kyai* had the students line up in a chain and pull each other's ears in rebuke so that he did not tire himself by punishing each of them individually. This was met by much hilarity from their fellow students. *Santri*, as a whole, are more alert and well prepared for a time, after such displays of discipline.

It should be noted that taking attendance and checking homework is a new innovation in *wetonan* education because greater discipline is required to compensate for the fact students now spend shorter periods of time in *pesantren* than in the past. This is true for *pesantren* that have added secular education, like An Nur and Tebu Ireng. However, this is also true for some *salaf pesantren* as well because prior to entering the *pesantren*, students are pursuing a general education for the six-year minimum (increased to nine in 1995) that the government requires.

Many Indonesian sources report that in *wetonan* anyone is free to attend as much or as little as they like (Arifin 1993:38; Syukri 1994: 64). This more open pattern is strongly reminiscent of what Makdisi reports in medieval study circles (*halqa*[5]) in the congregational mosques (1981:18). I observed this more open pattern at various *pesantren* during *Ramadan* and school vacations, when ordinary lessons are suspended, some regular students go home, and those who remain are joined by wandering students[6] for short courses on particular books. Most *ngaji* education is a variation of *wetonan*. Younger *santri* are taught by more senior *santri*, who are still studying under the *kyai*. In addition to *wetonan* and *sorogan*, Imron Arifin recognizes three other patterns of *pengajian*: *mushawara*, *muzakkirah*, and *majlis ta'lim* (1993:38–40).

The *mushawara* method involves the practice of Arabic conversation. Some modern *pesantren*, like Gontor, take this to the point of forbidding all other languages, except English, for use by *santri* (Adnan 1981). In many *pesantren* this practice is not required every day but only a few times a week, and it is often combined with the reading and giving of *khitabah* (sermons written by a famous *ulama*) (Syukri 1994:65). Advanced students will read these well-rehearsed sermons in village mosques during Friday prayers.

The *muzakkirah* method uses a discussion group. Imron Arifin recognizes two levels of *muzakkirah* (1993:39). The first level involves small groups of *santri* discussing specific religious problems and questions. Each group appoints a spokesperson to report back to the larger group and to the *kyai*. The second level is led by the *kyai*. In this setting, the findings of each group are reported and evaluated. For less advanced students, the two levels will be held in the same setting, so that any errors can be corrected immediately. More advanced students, however, may meet repeatedly without a teacher monitoring their discussion. They may occasionally report back to the *kyai*, but for the most part are self-policing, crosschecking their opinions against one another and coming to a consensus. Arifin argues that this is done primarily in Arabic (1993:39). In my observations of such group discussions, however, the texts were read in Arabic but the discussion that followed was conducted in Indonesian.

The last method of *pengajian* that Arifin recognizes, *majlis ta'lim*, is not really a method of education for *santri* in a *pesantren*. Although *santri* may attend, *majlis ta'lim* are geared toward the general public and cover basic religious advice (1993:67). A variation of this is routine *pengajian* for local religious leaders. I attended one of these with Kyai Masduqi, the vice-head (*wakil Rois Aim*) of the *Syuriah* (Islamic law council) for East Java NU. It not only uses a lecture format, but also has a question and answer segment. For the most part, the questions are submitted before the meeting, so that the *kyai* may research the answer. When I attended, one question concerned whether or not Muslims could participate in the celebration of the 50th Anniversary of Indonesian Independence. The answer was that as long as the means of celebration was in accordance with the laws of Islam, not only may Muslims participate but also it might even be considered required (*wajib*) because of the role the Islamic community, especially the *pesantren* world, played in that struggle.

Religious education involves studying texts, which include the Qur'an, hadith, and the classical texts that include commentaries on scripture, expositions on mysticism (*tasawuf, tariqa*), morality (*akhlak*), pedagogy, as well as texts on jurisprudence (*fikih*), doctrine (*akidah, usuladdin*), Arabic grammar (*nahwu, shorof, balaga*), and prayers and invocations (*dua, wirid, mujarrabat*) (van Bruinessen 1990:229). Martin van Bruinessen identified 900 different *kitab kuning* that are used in the *pesantren* world. Just over half of these works were written or translated by Southeast Asian *ulama*. Of the 500 works in Arabic 100 were written either by Southeast Asians or expatriate Arabs living in the region. Approximately 55 percent of the works were in Arabic, with the remaining works written in Indonesian languages, with those written in Malay having the largest representation (van Bruinessen 1990:229). He classifies the works into the following categories and

Table 3.1 Comparison of texts used at Tebu Ireng, An Nur, and in general

Subject	General usage (900 titles) (%)	Tebu Ireng usage (28 titles) (%)	An Nur usage (62 titles) (%)
Jurisprudence	20	10	20
Doctrine	17	7	12
Arabic grammar	12	43	30
Hadith collections	8	10	14
Mysticism	7	3	9
Morality	6	7	8
Collections of prayers and invocations	5	0	0
Texts in praise of prophets and saints	6	0	0

Source: Values for general usage (900 titles) taken from van Bruinessen (1990:229).

calculates the percentage each category represents out of the total number of texts in use. I have done likewise for two specific *pesantren* (see table 3.1).

For our purposes here, it is sufficient to summarize that the majority of the *kitab kuning* are commentaries within the established schools of thought. This is to say that the *fikih* texts are firmly Shafi'i in their orientation, the mystical texts are Ghazalian, and the theological texts are Asyarite. I explore the implications of these basic schools of thought in the next chapter and leave a nuanced examination of the fine differences between different texts to someone else.

By comparing the books used in a particular *pesantren* to the overall count made by van Bruinessen, we can see the focus of these *pesantren*. For instance, we would expect that in a *pesantren* known for mysticism, *kitab* on mysticism would make a larger percentage of the *kitab* used than in general. Further, *pesantren* that are focused on basic training will have a greater percentage of books in Arabic grammar, emphasizing the basic skills of reading and writing classical Arabic. Tebu Ireng and An Nur are of a similar sort; they both try to balance traditional *pesantren* education with government-recognized curricula. However, looking at these data, we can see that Tebu Ireng has a greater focus on the basics, mostly grammar. Further Tebu Ireng teaches more than 50 percent fewer texts than An Nur.

Pengalaman *and Moral Education*

Pengalaman, experience, is another component of *pesantren* education. *Pengalaman* may include specific training in the giving of sermons (*khotbah*) and other forms of public speaking. Sometimes senior *santri* are sent out to nearby villages to give the sermon at the Friday Prayers or to lead the prayers for a *slametan*. However *pengalaman* is most often concerned with moral education, that is, the application of the values learned in *ngaji*. According to Nafik of Al-Hikam, the worst human is one who has knowledge but no morality and hence it is important that knowledge and morality be taught together. Nafik offered the role of *pesantren* in the fight against Dutch colonialism as an example of what occurs when knowledge and morality are joined together.

Briefly, the moral values that are emphasized in *pesantren* include *Ahkwuya Islamiya* (Islamic brotherhood), *keikhlasan* (sincerity, unselfishness), *kesederhanan* (simple living), and *kemandirian* (self-sufficiency). Beyond these, *pesantren* aim to instill personal piety and a commitment to the Five Pillars: *shahada* (confession of faith), *salat* (five daily prayers), *zakah* (almsgiving), *puasa* (fasting during the month of *Ramadan*), and *hajj* (pilgrimage to Mecca, for those who can afford it).

Pesantren teachers stress that although a daytime *madrasah* can teach students about religion and morality, they cannot teach the students to be moral. Moral education in terms of moral behavior, takes experience. Hence, *pesantren*, as boarding schools, strive to create an environment in which the morals of religion can be practiced as well as studied. The students learn about them in *ngaji* and are then given the opportunity to practice them.

For example, although five daily prayers are required in Islam, doing them communally is not. However, communal prayers (*salat jamaah*) are seen as a better way to pray and are generally required in *pesantren*. In fact, a *pesantren* that does not make or enforce this requirement is considered by some to no longer be a real *pesantren*. Leaders say that this practice teaches brotherhood and community, values that the Islamic community needs to thrive; if once a week (on Fridays) defines a community, then gathering every time cements the bonds of said community. A senior *ustadh* at Al-Hikam said *salat jamaah* also teaches a model of leadership. If those in the back see that the *imam* (prayer leader) has made a mistake, they remind him of the correct form by saying "*Subhana Allah*" (praise be to God), not as a protest, but as a reminder. Another *ustadh* continued to explain that if the *imam* passes gas, and thereby breaks his *wudu* (ritual purity), he must step out of the way and allow someone else to take over. The prayers of the community are not invalidated by this, but rather protected. As a political model, it suggests an synergistic relationship between the leader and the led.

Other values, such as *ikhlas* (selflessness) and *kesederhanaan* (modest living) are taught by spartan and communal living arrangements.

In most *pesantren*, the *santri* sleep on the floor in a room that may hold up to 80 other students. A room that I would judge to be adequate for one or two students, houses six to eight; the more popular the *pesantren*, the more crowded the space. The meals are meager: rice and vegetables. Further, although there is an acknowledgment of personal property, in practice, property is communal. Simple things such as sandals are borrowed freely. Other items, if not in use, should be loaned if asked for. The *santri* who habitually refuses to loan his property will be sanctioned by his peers and sometimes by the *pesantren* staff. I was expected to follow these guidelines as well and I often found my tape recorder and camera missing. They were always returned later, the camera with all of its film used and with a request to have the film developed. For the *santri* who does not share, sanctions may include teasing or a stern reminder about Islamic brotherhood and the importance of being *ikhlas*.

In many ways, the details of the *pesantren* lifestyle have not changed much over time, however given changes of lifestyle and standard of living in the general population there is a greater gap between the two and hence the *pesantren* lifestyle is more ascetic. In other words, once the simple lifestyle was a matter of necessity, neither student nor *kyai* could afford more, but now it is a matter of choice. Kyai Baddrudin of An Nur told me that an ascetic lifestyle in the *pesantren* prepares the students for either prosperity or poverty. In the former, they will be compassionate; in the later, they will be content.

The value of *kemandirian* (self-sufficiency) is taught by having the *santri* take care of their own basic needs. The essential idea of this value (*mandiri*) is seen in a common joke. I was told repeatedly, in the presence of very young *santri* (6–7 years of age) that *mandiri*, the root of *kemandirian*, was an abbreviation for *mandi sendiri* (bathe on your own). Although this joke was always met with great hilarity, it communicates quite clearly, both to the young *santri* (who may still be used to bathing with older siblings) and to the foreign researcher, that taking care of oneself is an important value. In traditional *pesantren*, *mandiri* manifests itself in cooking arrangements; students cooked for themselves, or in small cooperative groups. Today, to regain time for *ngaji* lost to general education, many *pesantren* employ a cafeteria system. However, *santri* still learn self-sufficiency through doing their own washing, ironing, and housekeeping.

Other rules enforced in most *pesantren* regard the nonattendance of lessons or communal prayer, sneaking out of the compound, watching movies, theft, and other activities deemed to be against *pesantren* values. Most violations result in the *santri* receiving stern advice (*nasehat*). Repeated violations may bring more stern discipline. One *ustadh* suggested that the punishment for minor offenses such as watching movies might include beatings or even being ordered to do push-ups in sewage runoff. If the violation is greater, the student's hair will be shaved off, often just before a scheduled parent's day event, which will humiliate the *santri*. Offending students may also be sent home. Ultimately, the form and force of the discipline is at the *kyai*'s discretion.

Gus Isyom of Tebu Ireng argues that in order to plant values (*menamkan nilai*), instruction is not as important as setting a good example. In order to teach his *santri* the importance of *salat jamaah*, a *kyai* needs to lead the prayers, not always, but often. Gus Isyom's cousin, Abdurrahman Wahid (then chair of NU) agrees that the living example of the *kyai* is critical in teaching *santri*. In this regard, Abdurrahman points to his uncle Yusuf Hasyim, who never teaches classical texts but does teach his students the importance of science and technology, by his activities outside the *pesantren* that allow him to bring government ministers to visit (there is a place for helicopters to land next to the campus for this very purpose). However, he is just as concerned about the morality of his *santri* as was his father (Hasyim Asyari); whereas Hasyim Asyari was concerned over the impact of popular music, Yusuf Hasyim is concerned about the influence of television on *santri* and has curtailed viewing considerably.

Mustahin, also at Tebu Ireng, argued that like the Prophet, *kyai* should be examples to their students so that *pesantren* education is not only about religious knowledge but also about moral character. Mustahin said that the Prophet had something called the *Sufa* in which 400 followers lived together and witnessed the revelation of the Qur'an. They lived together in a religious community and studied from the Prophet. In this context, they were able to study not only religious knowledge, but also how the Prophet actualized his faith. In this regard a *kyai* must live in the *pondok* so that he can give his students an example of an Islamic lifestyle. If he does not provide this example, then the education is instruction (*pengajaran*) only and not true teaching (*pendidikan*). In this way, he suggests, the personality and the character of the *kyai* himself is a central part of *pesantren* education, a point I return to later in the chapter. Some *pesantren* directors are so involved in politics that they are rarely in their *pondok*.

Gus Isyom suggested that mysticism is central in moral education. He expounded that in Islam there is a "triangle" of major "sciences" (*ilmu*): *tauhid* (theology; especially as regards the nature of Allah), *fikih* (religious law), and *tasawuf* (mysticism) (see figure 3.1). Each of these sciences makes different contributions. *Tauhid* establishes the basis of faith. However because faith is not enough and needs good

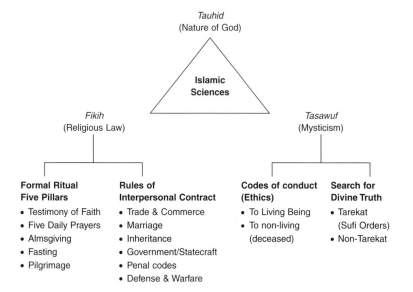

Figure 3.1 Triangle of Islamic *Ilmu* (Sciences) (cf. Muhaimin 1995)

works (*amal*) to actualize it, *fikih* provides the believers with guidelines on how to live right and perform good works. And yet because good works are empty if the motivation is impure, *tasawuf* (mysticism) instills moral and ethical values in the believers. The centrality of *tasawuf* in learning ethics and leading a moral life was among the reasons why Hasyim Muzadi arranged to have Syehk Abdurrahman visit Al-Hikam as related in the first chapter.

The association of Sufism and ethics may be traced to a single highly influential Islamic thinker, Imam Al-Ghazali. Al-Ghazali is regarded by some as the *mujaddid* (A., reformer) of the fifth century of the Islamic era in fulfillment of the prophesy found in hadith, "that at the beginning of every century God will send someone to revive and revitalize the faith of the Islamic community"; some even accept him as the greatest religious authority after the Prophet (Abdul Quasem 1975:11). Al-Ghazali is famous for his sober mysticism, which balanced theology and *tasawuf* (mysticism) and his extensive works on ethics. It is through the use and study of Al-Ghazali's works that many in the *pesantren* world associate mysticism and ethics.

Sekolah *and General Education*

Sekolah (school) is not found at all *pesantren*. *Sekolah* usually includes one of the two basic government recognized curricula one secular, called here the National System, and the other religious, called here the *madrasah* system. These are now in the process of being combined. *Pesantren* may have neither, either, or both types of schools within their grounds.

The first system was under the Department of Education and Culture and was largely secular. However, partly because of the insistence of Wahid Hasyim, an early minister of Religious Affairs and a famous *kyai*, this education includes two hours each week of religious instruction. Ideally students are taught only their own religion, but classrooms with students of multiple religions create problems. For example, students who attend a school run by members of another religion are not required to attend the religion class. Since a Catholic school will not offer Islamic education, and vice versa, these students may receive no religious instruction at all. The schools under this system were called *Sekolah Dasar* or SD (elementary school), *Sekolah Menengah Pertama* or SMP (junior high school; lit. first middle school), and *Sekolah Menengah Atas* or SMA (high school; lit. upper middle school). There are also technical and vocational high schools.

The second system was the *madrasah*[7] system under the Ministry of Religious Affairs. The state *madrasah* system was originally founded because many parents would not send their children to the largely secular state schools. In 1994, Zamaksyari Dhofier, then head of the Directorate of Islamic Institutes Development, stated that 20–30 percent of all Indonesian students attend *madrasah*. He further stated that in some areas this was as high as 35–40 percent because *madrasah* are often associated with *pesantren*, and some parents are more likely to trust a *kyai* with their child's education than they will a government teacher (personal communication). Some parents will only allow their children to attend school if it is at a *madrasah*. The schools under this system were called *Madrasah Ibditayyah* or MI (elementary school), *Madrasah Tsanawiyah* or MTs (junior high school), and *Madrasah Aliyah* or MA (high school). When Dhofier conducted his dissertation research in 1977–1978, the curriculum at *madrasah* schools was 70 percent religious and 30 percent general subjects. In 1984, the curriculum was restructured so that these percentages were reversed. Ghaffar Rahman, who became the Head of NU's school-based education program (*Lembaga Ma'arif*) in 1995, but has also served this organization on various levels since 1968, argues that even before 1984, there were some who felt that the religious education was not sufficient. They would address this problem by adding hours to the day, shortening each class period, to add an additional class period, and other acts of bureaucratic sleight-of-hand. With the severe reduction in 1984, this feeling increased.

The final blow came in 1994, when the percentage of religious education in *madrasah*, for most students, was reduced to 11 percent (see table 3.1).

The curriculum introduced in 1994, *Kurikulum 94*, begins to combine these two systems at the high-school level. Eventually both systems will be under the Department of Education and Culture, at least for the majority of the curriculum; religious education will still be under the control of the Ministry of Religious Affairs. The *Sekolah Mengenah Atas* is being renamed *Sekolah Mengenah Umum* (general middle school). The *Madrasah Aliyah* is being divided into two pro- grams, the *Madrasah Aliyah Keagaman* or MAK (the upper religious school), and *Sekolah Mengenah Umum yang berciri khas Islam* or SMU Islam (general middle school with an Islamic character). The MAK program is more than 70 percent religious and uses Arabic as the language of instruction. The SMU Islam has only 5 out of 45 credits in religion, or approximately 11 percent compared to the SMU that has two credits of religion. Whereas the religion lessons at the SMU are quite general, the courses in the SMU Islam are more specific: *fikih* (jurisprudence), *akidah-ahklaq* (doctrine and morality), Qur'an, and hadith.

Abdurrahman Wahid, then general chairman of NU, argues that the increasing percentage of general subjects in the *madrasah* curricu- lum is part of the government's program to improve human resources. Most *kyai* I spoke with agreed with this assessment and did not see the continual reduction of religious curriculum in the *madrasah* as an attack on Islam, or at least they would not say so. However, many feel that the reduction left too few hours for religious training and hence they seek ways to correct this. Some already had methods in place because they felt that the previous curriculum was lacking in religious education. This illustrates that when a *pesantren* adopts government regulated education they are subject to some outside control.

All private schools are ranked by the appropriate government ministry. Schools under the Department of Education and Culture may be given one of three rankings: *terdaftar* (registered), *diakui* (recognized), and *disamakan* (equalized). Schools in the *madrasah* system are given one of two rankings: *terdaftar* and *diakui*. No private *madrasah* is ever given the status *disamakan* because the Department of Religious Affairs is not empowered to do so. It is beyond the scope of the present work to fully analyze the educational effectiveness of secular education in *pesantren*. However, it should be noted that *pesantren* vary widely in the degree to which they stress the impor- tance of secular education. In one *pesantren* studied, the *pesantren* staff were very concerned if students missed communal prayers but did nothing if they saw a student skipping class. In another *pesantren* studied, the exact opposite was true.

Kursus *and* Ketrampilan

Traditionally students did not pay for their education or lodging but worked for the *kyai* in exchange for their expenses, a pattern found elsewhere in the Islamic world (Mottahedeh 1985:95). In the course of this work, they may have gained some skills that they might put to use after they returned home. As of late, it has become vogue for *pesantren* to offer extra courses (*kursus*), with English and computer skills being most popular, and job skills, such as chauffeuring, automobile repair, sewing, small business management, and welding. In part, this is in response to government programs encouraging the improvement of human resources. Often, this training still takes the form of working in exchange for educational expenses. Because the addition of general education has meant less hours in the day for religious study, it is now more common for students, or their parents, to pay directly for the student's expenses. This has sometimes meant that only those who cannot pay receive vocational training. On the other hand, *kursus* may be offered but on an infrequent basis.

Even if *kursus* are occasional and skills training (*ketrampilan*) minimal, or limited to work related training, these activities become important in the social construction of what a *pesantren* is, or should be. Between the government's *Meningkatkan Kwalitas Sumber Daya Manusia* (Improve the Quality of Human Resources) Campaign and the very real need for graduates to earn an income, a *pesantren* that does not address these issues, or at least claim to, will quickly become unpopular. Still, religious education is paramount; if students and parents were primarily interested in job skills, there are many schools that could provide said skills, and do a better job of it. Often, the desired end product is a person who has traditional religious beliefs and values along with just enough education and skills to make a living.

Curriculum and Types of Pesantren

In Indonesian discourse, how a *pesantren* engages these areas leads to one of three labels: *salaf*[8] (A., traditional), *khalaf* (A., modern), and *terpadu* (I., mixed). *Salaf pesantren* have only religious education and character development. *Salaf pesantren* best preserve the teaching of classical texts as essential education. *Khalaf pesantren* are characterized by religious education conducted exclusively in Indonesian, and

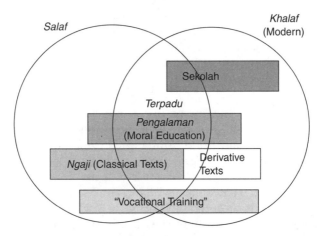

Figure 3.2   Types of *pesantren* and types of curriculum

by the importance placed on general education and skills training. The less an institution emphasizes religious education and character development, the less likely it is to be considered a true *pesantren*. The most traditional *pesantren* tend to limit the innovations used in the teaching of this curriculum. According to IAIN faculty and students, *salaf pesantren* are known for the best religious education, or at least the most traditional, and modern *pesantren* are best known for their capacity to teach Arabic and English. However, some argue that *salaf* graduates are not as well prepared to deal with broader social issues facing a changing society. Therefore they are not as readily accepted by the masses as graduates of modern *pesantren*. However, this assertion is contested and cannot be accepted at face value. Most *pesantren* today are labeled mixed because they engage some combination of all four types of curriculum. Even so, there is considerable variation within this category (see figure 3.2).

Ideal types of *pesantren* emerged from individual and group interviews. First and foremost, *salaf pesantren* leadership is concentrated in the hands of one man, who holds complete authority. It is often said that a *kyai* is a "little king" (*raja kecil*). Further, the *santri* treat the *kyai* with much respect, never looking him in the eye, especially when spoken to. When the *kyai* passes, students should make way for him; some will even crouch down and bow their heads. When the founder dies and his sons succeed him, the model of a single authority often leads to a leadership crises and to the fragmentation of the *pesantren*. Yusuf Hasyim, headmaster of Tebu Ireng, suggests that such dynastic in-fighting has caused the downfall of many large and famous *pesantren*.

The education in *salaf pesantren* is exclusively religious in nature, that is just *ngaji* and *pengalaman*. Classical *kitab* are usually studied using the *wetonan* method described above. According to IAIN students, this largely passive pedagogy does not allow for the development of real thinking (read: critical, historical thinking), which is a claim about their being unsuitable to aid in Indonesian development. Although the graduates of *salaf pesantren* sometimes received diplomas, they are not recognized by the government. To attend schools of higher learning, graduates of *salaf pesantren* have to take an equivalency test.

In general, *salaf pesantren* do not want interference or influence from the outside. Therefore they tend to refuse government assistance. The lifestyle of the *santri* underscores this opposition to outside influences. Magazines, newspapers, radio, cassette tapes, and television are not allowed. Further, students are not allowed to leave the *pondok*; their clothing and personal needs are met by a store inside the *pondok*.

Pondok Lirboyo in Kediri is held by many to be a good, if not the best example of a *salaf pesantren*. My short visits to Lirboyo showed the centrality of mystical practice and personal prowess in martial arts. One visit was admittedly during an annual *pencak silat* (martial arts) exhibition. One central feature of this *pesantren* was the graves of the founders, which had large numbers of students praying beside them at all times. In order to attend college, Lirboyo graduates must take an equivalency exam. The three Lirboyo graduates that I interviewed were all sons of *kyai*. Two of these three added additional education to their Lirboyo training. The son of one regional NU functionary took the equivalency exam and is pursuing a degree at IAIN Malang. He said with dismay that most of his IAIN classmates and many of his teachers cannot read the classical texts. In order to have both government credentials and a quality religious education he has had to endure a double training process. Gus Isyom of Tebu Ireng earned a *Sarjana Hukum* (B.A. in secular law) while attending and teaching at Lirboyo. His knowledge of both secular and religious law causes many to expect great things from this young man (late twenties in 1994).

Modern *pesantren* are said to use Western pedagogical theory and practice. These are better known for their training in Arabic and English. However, the religious education is not as strong because it is based on derivative texts and the students do not study *kitab kuning*. These are said to offer a government recognized diploma, although ironically the prototypical modern *pesantren*, Gontor, does not. The leadership at modern *pesantren* is said to be more open and democratic than that found at *salaf pesantren*. The leadership is based on the knowledge of the *kyai* and not on other factors such as family heritage or charisma. Often there is more than one *kyai* and sometimes a triumvirate that is elected by a governing board, that is the case at Gontor, the prototypical modern *pesantren*.

The typology given above is far from static, even if it reduces a dynamic process to a static model. Javanese, both from within and without the *pesantren* community, have created these typologies as a way of getting a handle on the changes taking places. However, the fact that a majority of *pesantren* could be placed in the *terpadu* category speaks of the extent of the changes taking place. There is a great deal of variation within this *terpadu* category as the comparison of curriculum between An Nur and Tebu Ireng suggested. The very term for the middle category, *terpadu* or mixed, suggests strongly that changes are taking place. The fact that most *pesantren* would classify themselves as *terpadu*, means that it is highly desirable to be seen as a school that is balancing traditional religious education and modern educational needs.

Although these typologies are discussed in Javanese society as points along a continuum, I favor a dialectical model. The pressures for *salaf* and *khalaf* education have created something entirely new, the *terpadu pesantren*. And although the term mixed does not imply something new, the concern for balance between basic knowledge and religious knowledge is indeed novel. These novel attempts to deal with the modern world have sparked considerable amount of debate. This debate will create new kinds of institutions. For example, since many *pesantren* can no longer graduate people who are able to become *kyai*, many in the *pesantren* community have started discussing the possibility of advanced *pesantren*. If contemporary *pesantren* are becoming more like middle and high schools because of their association with secular versions of those schools, then the advanced *pesantren* would be a return to the original seminary function of these schools.

Discussion

The discourse about *pesantren* history is part of the process of inventing tradition. The nature of the *pesantren* tradition is contested, however, throughout the discourse there are some common themes.

The first theme is that *pesantren* are Indonesian. The claim that *pesantren* have their basis in pre-Islamic local institutions is ultimately a claim that *pesantren* are local and not an export from a foreign culture. Even those who dispute a Hindu–Buddhist origin for *pesantren*, argue that *pesantren* are somehow unique to Indonesia, particularly in terms of the combination of Islamic scholarship and Islamic mysticism. Further, *pesantren* are Indonesian because of their role in the revolution. In other words, there would be no Indonesia without the war efforts of *pesantren* people. Therefore, *pesantren* have a vested interest in the future of Indonesia and hence will take up the education challenge to prepare future cadres of leaders and followers. Finally claiming that the Wali Songo were the founders of the *pesantren* tradition ties local tradition to Islam tradition.

The second theme is that *pesantren* practice teaches the proper form of Islam. The claim that *kyai* are *pewaris Nabi* (inheritors of the mantle of the Prophet) makes clear claims about the correctness of *pesantren* practice. The emphasis on classical texts also defines the tradition in a particular way: one in which people are allowed to adapt Islam to the modern world, but not without the wisdom of centuries of Muslim scholars before them.

A central part of the strategies used by the *pesantren* community to engage modernity is education. By adopting government-recognized curriculum, these schools are aiming to prepare their students to participate in modern Indonesia and wider global economic, cultural, and political processes. However, by placing these kinds of educational opportunities within the context of a *pesantren*, they are making particular claim about the nature of the modern world for which they are preparing their students. Namely, that modernity is in need of tempering by the traditions of the *pesantren* community.

Some of these curricular changes have to do with maintaining influence over society. Since it is the head of a school, rather than the functionary at a place of worship, who leads the Javanese Islamic community, maintaining *pesantren* as viable schools is critical to maintain classical Islam in Java. Without large numbers of students who will have lifelong allegiances to a *kyai*, the headmaster's ability to influence society is greatly reduced.

In this context, it may seem that *salaf pesantren* are reactionary and reject this accommodation with modernity and at some level this may be so. However, several *salaf pesantren* I visited encourage students to complete the government-required minimum nine years of recognized education prior to beginning studies at the *pesantren*. Further, several *ulama* families have adopted the strategy of a double education for their children (primarily sons) in which they complete government-recognized education through the college level and then seek a more traditional *pesantren* education at a *salaf pesantren*. In this way *salaf pesantren* become part of the overall strategy of the *pesantren* community to maintain tradition while adapting to modernity.

In order to clearly define tradition and shape modernity, *pesantren* have to maintain certain aspects of traditional *pesantren* education. Practices like ascetic lifestyles, and communal prayers as well as the emphasis on classical Islamic texts are ways to define the *pesantren* tradition. In the next chapter, we will explore additional ways in which this tradition is defined.

Chapter 4

Beyond Education

One of the most lively discussions I remember observing was about toothpaste. *Kyai* from all over East Java had gathered to debate this and other issues, but this one seemed to take center stage. The toothpaste in question was called Siwak-F that contains both fluoride and powdered *siwak* wood. The question put forward was whether Siwak-F had the same ritual and legal status as using a stick of *siwak*; that is, whether or not the modern practice of brushing teeth could purify one's mouth for prayer and Qur'an recitation the same as the Prophet's practice of using the wood of a particular tree. It had already been established that brushing one's teeth is a good substitute when one does not have the special wood; the debate here was whether using the toothpaste with the ground wood in it would have the same ritual/legal status as using the wood itself. All the conference participants had been given a free sample tube. The manufacturer was clearly interested in the outcome of this debate. Several prepared papers were passed around and every *kyai* who wished to speak on the issue did so. The *kyai* who had gathered from all over East Java were of many opinions and seemingly no conclusion was met at this particular gathering. However, the case illustrates both the process for debating such issues and the concern for balancing modern needs (i.e., good dental care) and ritual purity.

The discourse in the *pesantren* world is concerned with far more than education. Ultimately it is about identity: how to engage modernity at the level of the nation *and* the global community *and* be a good Muslim. Abdurrahman Wahid said that in a seminar, he was asked to speak about the relationship between universal Islam and nationalism. This issue is one not directly related to religious education, however, it is an issue that *pesantren* have to address. He told me,

> I spoke of the problem as something that is both universal and exclusive
> to Muslims. The relationship between the two is problematic because
> of the relationship between Indonesian Nationalism and Islam.
> Indonesian nationalism is problematic because it was born of compro-
> mise; the requirement of *shariah* law was erased from the Jakarta
> Charter. At that point the relationship between the two became
> unclear. Is *shariah* codified or not? Does the nation enforce *shariah*
> law (in marriage, etc.) or not? Islamic universalism is also problematic.
> It claims to be universal, but it only for Muslims. This problematic rela-
> tionship has given rise to a moral crisis. There is no set of values that are
> held to apply to all Indonesians.

There are many levels of identity that *pesantren* people are trying to
resolve. First, is the question of traditional identity. What does it mean
to be a classical Muslim? To do this they tap into larger Muslim
discourses and symbol set. They use contested symbols and forward
a position that holds their traditions as the best way to be Muslim.
Second, there is the related issue of the appropriate relationship
between Indonesian Islam and Arabian Islam. The question asked is
to what extent must Indonesians be culturally Arab in order to be
good Muslims. Finally, there is a question of the appropriate relationship with the Indonesian State.

Defining Traditional Identity

The broadest-level identity of the *pesantren* world is *ahlus sunna wal
jamaah*. Simply *ahlus sunna wal jamaah*, abbreviated "ASWAJA" in
Indonesian discourse, means "the followers of the Prophet and the
consensus of the *ulama*." There is a well-established hadith in which
the Prophet predicted, "my Umma will split up into seventy-three
sects: seventy-two will be in the Fire, and one in the Garden"[1] (Hasan
1994). Hence, ahlus sunna is defined as the only sect keeping the
original teachings of Islam. At this most basic level of definition the
term is synonymous with Sunni Islam.[2] However, in Javanese discourse, this tradition is defined by distinguishing features in three
areas: theology, jurisprudence, and mysticism (Dhofier 1980b:297).
In Java, the theological aspect is largely undisputed because most
Indonesians subscribe to the *Asyari* theological school. It is these
particular approaches to jurisprudence and mysticism that shape the
meaning of ahlus sunna in the *pesantren* world.

However, in the study population there is an overriding sense of
being Muslim, and being Muslims of a particular type, often glossed

as *ahlus sunna wal jamaah*. *Ahlus sunna wal jamaah* means People of the Way of the Prophet. It is a slippery label of identity. In Java, it is most often associated with *Nahdlatul Ulama* (NU), which has used this term as a political and religious slogan (Federspiel 1996:200). When asked who practices ASWAJA, most NU people affirm that they do and imply that only they do. However, few are willing to state bluntly that Muhammadiyah, the other major Islamic organization in Indonesia, is not *ahlus sunna wal jamaah*; such a statement would place Muhammadiyah outside the fold of true Islam. Given the strong ethic of Islamic unity, such statements are rarely made directly. There is a well-known hadith that states, "when a Muslim calls a Muslim an unbeliever, then one of them is an unbeliever" (as cited in Vikør 1995:4), which means that if one is wrong in an accusation of unbelief, then one is judged to be an unbeliever.

Although *pesantren* people will not state that such and such a group is not ASWAJA, they will state that those with certain characteristics, which the knowing audience associates with the same groups just affirmed as being ASWAJA, are not ASWAJA. For example, Kyai Djauhari argued in a public lecture that Syehk Ahmad ibni Tamiyah, the thirteenth-century founder of Islamic "modernism," and those who later took up his teachings including Syaik Muhammad bin Abdul Wahab and his followers (the Wahabis) fall outside of *ahlus sunna wal jamaah*, which in the Indonesian context implicates Muhammadiyah. However Djauhari quickly stated,

> In the matter of understanding between the followers of NU, Muhammadiyah, Persis, Perti Al-Washiyah, they all may be placed within the container of *ahlus sunna wal jamaah*.

Djauhari placed all Indonesian Muslims within the identity of ASWAJA, while at the same time defining those who follow Ibn Tamiyah or Abdul Wahab (as some of these groups do) as outside of ASWAJA. Muhammadiyah people also see themselves as having a better grasp on the truth of Islam, and argue that Muhammadiyah is *ahlus sunna wal jamaah* because of its insistence on basing all beliefs and practices on the Qur'an and strong hadith (Kamal et al. 1988:156).

Features of ASWAJA

An important feature of ASWAJA, even if not disputed in Java, is following the teachings of Imam Abu Hasan Al-Asyari (b. 873, d. 935 C.E.) and Imam Abu Mansur Al-Maturidi (d. 944 C.E.) in matters of Islamic theology (Dhofier 1980b:297–298). Although there is some variation in their theology, they are referred to as the joint founders of Sunnite rational theology (Watt 1974:187–189); they both advocated a position in which revelation is more fundamental as a source of truth and in which reason is used to confirm revelation (Ali 1961:265; Hye 1961:231). For the most part, both NU and Muhammadiyah follow this school and hence differ not in theology but rather in terms of jurisprudence and mysticism; this aspect of ASWAJA identity is not generally contested in Indonesian Islamic discourse.

In regard to jurisprudence, being *ahlus sunna wal jamaah*, according to *pesantren* people, means adhering to one of the four *madhab* (ritual/legal school) in the matters of Islamic jurisprudence. The four *madhab* include the Hanafi, Hanbali, Shafi'i, and Maliki schools. The Hanafi school was founded in Iraq, the Hanbali school in Baghdad, the Shafi'i school in Egypt, and the Maliki school in Egypt, and North Africa (Lapidus 1988:164). In Java, most *kyai* are strict followers of the Shafi'i school, which was founded by Muhammad ibn-Idris Al-Shafi'i (b. 767, d. 820 C.E.). In the Shafi'i *madhab*, the roots of law are: the Qur'an, the hadith, the consensus of the *ulama* (*ijma*) and analogical reasoning (*qiyas*). Besides Java, the Shafi'i *madhab* is also predominant elsewhere in the Malay Archipelago, as well as in South India, Egypt, Syria, the Hijaz, South Arabia, parts of the Persian Gulf, East Africa, Daghistan, and parts of Central Asia.

The insistence on following one of the four law schools is particularly meaningful in Javanese/Indonesian discourse. Muhammadiyah claims that the basis of Islamic law should be Qur'an and hadith only. By so doing, they reject all historical developments in Islam including the *madhab* (Peacock 1978:35n, 107). I was told repeatedly by *pesantren* people, that Muhammadiyah was its own *madhab* and did not follow one of the four established law schools. This implies that Muhammadiyah might be outside the fold of *ahlus sunna wal jamaah*. However, given the strong incentives for tolerance and Islamic unity, such as an ideal of Islamic brotherhood and firmly enforced government regulations against any divisiveness based on race, ethnicity, or religion, it was rare for there to be statements that directly place other groups outside of the fold.

The final defining feature of ASWAJA in Java given by Dhofier is to follow the teachings of Imam Abu Qasim Al-Junaidi (d. 910 C.E.) in matters of Sufism (Islamic mysticism) (Dhofier 1980b:298). Al-Junaidi is recognized as the founder of sober Sufism. Al-Junaidi was able to win the approval of orthodoxy as relatively safe from *shirk* (polytheism) and other heresies. Because of this, he is included in the genealogies of most Sufi brotherhoods. Therefore to say that the *pesantren* world follows Al-Junaidi is to say little more than they practice Sufism that has the stamp of orthodoxy. However, the mysticism of the *pesantren* world can be further defined.

ASWAJA Mysticism

Kyai A. Wahid Zani, the head of RMI, a *pesantren*-focused organization within NU, argues that in addition to the mysticism of Al-Junaidi, *ahlus sunna wal jamaah* also follows the mystical teachings of Imam Al-Ghazali (Zaini 1994:51). The mysticism of Al-Ghazali is "a sober kind of mysticism carefully eschewing all kinds of pantheistic extravagances . . ." (Sheikh 1961:617). Al-Ghazali is called a "middle-roader" and in this regard strove to make mysticism orthodox as well as to make orthodoxy mystical. He held that the mystical element of religion as its most vital part, without which religious life is not possible (Sheikh 1961:617). Although his life history meanders through various religious approaches, in his later life he thought that theology and *tasawuf* (mysticism) were equally important, an opinion oft repeated in the *pesantren* world, and one reflected in the dual function of *pesantren* as centers of learning and centers of mysticism.

As *ahlus sunna wal jamaah* is practiced in East Java (at least), its mysticism can be defined even further. At a *haliqoh* (religious seminar from the Arabic *halqa*, study circle) at which several prominent *kyai* were present, it was agreed that being ASWAJA meant observing *manakib* (reading the histories of prophets and saints). When Kyai Tholchah Hasan pointed out that Shi'i Muslims observe *manakib*, another *kyai* pointed out that it was the reading of the *manakib* of Syehk Abdul-Qadir Gilani that was a defining feature of ASWAJA. The attendees were in general agreement about this point.

Manakib are recited in Arabic and although there is an occasional translation and commentary in Indonesian (e.g., Abu Amar 1989), such translations are not common. I witnessed *manakib* on several occasions, the first of which was at An Nur. Here it was observed by the senior *santri* on the eleventh day of the Javanese month. In my notes I observed,

> The manakib was held in a classroom. The chairs and tables were pushed to the back. About twenty-eight people attended. Noticeably absent was Kyai Badruddin, who was reviewing the video tape I shot of the annual memorial service (*haul*) held for his father, only a few days before. Everyone sat on the floor in a semi-circle. Someone sat in front of the group and lead the chanting. They chanted for about thirty minutes the history of Shaikh Abdul-Qadir Gilani, who Akbari identified as the founder of the Qadiriya sufi order. After the chanting ended, the

leader stepped out and large trays of food were brought in. It was white rice, noodles, vegetables, chicken, eggs. We sat in groups of four or five around the large trays and ate "like the Prophet" from a common plate with our right hands. Afterward, one tray of leftovers was given to the *santri* who facilitated the *manakib*.

Manakib are performed to give thanks for knowledge gained in *pesantren* studies and to seek spiritual power (*barakah*). In *manakib*, the participants use Syehk Abdul-Qadir Giliani as an intermediary. Qadir Gilani is the founder of Qadiriyya, a Sufi brotherhood bearing his name. The analogy used to explain this is that if you wanted something from your boss, you might go through a friend to get it, especially if your friend was on better terms with boss.

Manakib may be read on a routine basis or on special occasions to guarantee blessing and safety, for example, before *santri* go home for the end of Ramadan holidays (*Eid al Fitri*), and in this regard closely resemble *slametan* (cf. Woodward 1988). In fact, there are three closely related rituals whose names are used somewhat interchangeably: *slametan, haul,* and *manakib*. All three rituals involve the reciting of sacred texts and the sharing of a communal meal. They all confer blessing and safety on the participants; however, the *slametan* and the *manakib* may be performed exclusively with this goal. The *haul* is specifically conducted to remember, and to confer merit upon, a deceased *kyai*; some *slametan* are conducted to remember, and to confer merit upon, a deceased family member. In fact, *haul* are sometimes referred to as a kind of *slametan*. *Manakib*, in invoking the memory of a great saint, are similar to *haul*, and are also sometimes referred to as *slametan*.

Manakib can invoke any major Islamic figure. However, in East Java, the most commonly, if not exclusively, invoked figure is Abdul-Qadir Gilani (1077–1166 C.E.), who is called the captain of the saints and was the founder of the Qodiriyah brotherhood. Abu Amar argues that *Manakib* Syehk Abdul-Qadir Gilani was first taught to the people of Java and Nusantara (Indonesia) by the Wali Songo (the nine saints who brought Islam). Some of the impact of this text can be seen in two Javanese localizations of Islam; specifically, echoes from Gilani's hagiography are found in the life-stories of local Islamic figures. The first case concerns Sunan Kalijaga, perhaps the most famous of the Wali Songo, and certainly the most Javanese. He was a son of nobility, a *priyayi*, and he promoted *wayang* and *gamelan* (as a vehicle for spreading Islam according to my informants). However, there is at least one parallel between the story of Kalidjaga and that of Syehk Abdul-Qadir Giliani. Both men were ordered by their respective teacher to sit in a particular place and meditate until they returned. Geertz reports that a number of his informants said that the period of Kalijaga's wait was anywhere from 10 to 40 years (1968:28). According to a popular Indonesian language rendition of Kalijaga's tale, he waited just three years (Kafanjanji n.d.:97–98), the same amount of time as Abdul-Qadir Giliani wait for his teacher:

> Khidlir said to Shaikh *Abdul-Qadir* Gilani, "Sit here." So he sat down in the place pointed out by his teacher. And he stayed there for three years and once each year, Khidlir visited him and said, "Don't move until I return." (Abur Amar 1989:73)

Abdul-Qadir Gilani's hagiography has been used as a model for other Islamic figures in Java. Ann Kumar's *The Diary of a Javanese Muslim* (1985) concerns a certain Mas Rahmat who potrayed himself as a leading *ulama*, an accomplished mystic and as a source of *barakah*. About this text, Woodward states, "Like Abdul-Qadir Gilani, Mas Rahmat encounters *jin*, daring to meditate in caves and forests of which others are afraid" (Woodward 1989:109, see also Kumar 1985:80).

Although not everyone in the *pesantren* world belongs to the Qodiriyah *tarekat*, its teachings are important. Abdul Gani stated that although they have studied the teachings of Qodiryina at An Nur, in the 11 years he has been there they have not studied the teachings of any other Sufi order. This is significant because An Nur does not officially follow any particular order.

ASWAJA and the Discourse about Modernity and Tradition

In summary, the defining features of ASWAJA in East Java are the adherence to one of the four established law schools, following the *Asyarite kalam* (theological school), and a sober approach to mysticism, perhaps following the specific teachings of the Qodiriyah *tarekat* (brotherhood). Now that we have defined the basic characteristics of ASWAJA, we can examine some of the debates and discussions about it. I will start with an examination of a *haliqoh* (A; *halqa*, learning circle), or seminar, held at *pondok* Pesantren Salafiyah Shafi'i yah in Sukorejo, East Java. The seminar included several prominent East Java *kyai* and was an NU-sponsored event. The subject was the preservation and promotion of ASWAJA in Indonesian society.

Although the seminar was about the preservation of traditional values, many aspects of it were quite modern: the choice of speakers, the language used, and the format. The speakers at this seminar were KH Muhammad Tidjani Djauhari, from Pesantren Al-Amien, Prenduan Madura, a modern *pesantren*, and KH Tholchah Hasan, the rector of Universitas Islam, Malang, an NU-oriented college. Neither of these men could be considered *salaf kyai*, given that they are leaders of modern educational institutions. Kyai Tholchah is however, an alumnus of both Tebu Ireng and Tambak Beras, from a time when the religious credentials of these *pesantren* were still undisputed. Many of the other participants were more traditional *kyai* including at least one that was strongly *salaf*. Further, the language used was Indonesian, which placed this event within a national frame. Finally, the format used was a contemporary scholarly one, in which two papers were presented and then discussed. A more traditional format would have taken the form of a collaborative discussion (*musyawarah*) in which all participants would be equals; the time allotments alone (the speakers had 45 minutes and the commentators each had 5) indicate a more hierarchical structure to this event. The content and rhetorical style of the papers, especially Kyai Tholchah's were modern Indonesian, invoking empirical evidence rather than classical texts.

The seminar was concerned with how to preserve and promote ASWAJA and in this regard addressed both the challenges and threats to ASWAJA as well as the means to preserve and promote it. The challenges included various Islamic heresies including Mu'tazilah, Shi'i, and Wahabiya. Although it is beyond the scope of the current work to detail all these opposing Islamic theologies, a few are worth mentioning. The inclusion of Wahabism as a sect of Islam outside of the true believers, *ahlus sunna wal jamaah*, a claim confirmed by the prominent *ulama* writer Siradjuddin Abbas (1969:309–330), shapes the identity of ASWAJA in specific ways. Wahabism is the sect that now controls Saudi Arabia and Mecca. Therefore claiming that they are outside of ASWAJA, combined with a general disdain in the *pesantren* world for *ke-arab-araban* (linking Islam with Arabic culture), makes the claim that Javanese (or Indonesian) Islam is true Islam and need not imitate the Middle East.

Kyai Tholchah observed that in order to preserve and promote ASWAJA, they should not be concerned just with facing challenges from within Islam but more with the challenges of globalization.

> Preserving and promoting ASWAJA does not mean facing the challenge of Shi'i, but rather facing the reality of a wider social and cultural environment and the growth of dynamic, high quality science and technology.

Globalization rather than Islamic heresies is the threat with which both speakers were more concerned. Kyai Djauhari identified the following threats to ASWAJA:

a. A lifestyle that is more concerned with status and social standing without anticipating the implications of the emerging era. This is even more damaging because the lifestyles of our brothers and sisters are materialistic, egoistic, and, in general, unIslamic.
b. The impact of globalization includes the infiltration of culture and the destruction of thought patterns in the middle of our own society. Our defenses against these impacts are weak and alternatives are not yet ready.

   What is clear is the influence of films and pictures that show much sex and violence. According to experts and researchers, such films are responsible for much of the crime and amoral behavior lately.
c. The loss of values and norms caused by many causes, so that the observation of our religious teachings slackens. More dangerous still is if our enemies' efforts to infiltrate our society and insert their values, which are clearly in contradiction to the teachings of Islam, succeed and thereby shape a generation that is non-Islamic. Examples include secularism, materialism, hedonism, individualism, and the like.
d. The fragility of the brotherhood and solidarity within a single religious community (Islam), to the point that some of the arguments become fodder for the news media.

Kyai Djauhari's concerns focused mostly on the break down of morality in Indonesian society. In part, he blamed the Islamic community for not being unified. This in part explains why he defined ASWAJA to include all Indonesian Muslims. Although he did place some of the blame on the Islamic community, he placed most of the blame on outside forces that he did not define beyond linking them to scandalous films and various "isms" (secularism, materialism, hedonism, and individualism). He attributed these influences to globalization.

One of the attendees and appointed commentators, KH Drs. Muchidin Suwondo, argued,

> We cannot always complain. We cannot blame America; blame Japan. We cannot forbid them either. Neither can we forbid our own (Indonesian) television.

He continued to address a concern I heard oft repeated. Many in the *pesantren* world believe that the American film and television industry is purposefully trying to erode traditional values in Indonesia. In part, this belief is supported by the fact that the United States made a trade agreement with Indonesia, in which the importation of Indonesian textiles into the United States is contingent on easier access for American film exports to Indonesia (Barber 1995:91). It is interesting that Kyai Suwondo argues that they cannot affect Indonesian television even though it is censored by a government agency. In effect, he was saying that in the struggle to define Indonesian society as a Muslim society, they needed to avoid any direct conflict with the Suharto government.

Kyai Tholchah was also concerned with globalization and the breakdown of identity. He argued,

> Life these days is multidimensional and complexity has invaded all sectors of human existence, including economics, politics, social interaction, culture, and religion. The information boom supported by high tech development has already torn national boundaries, penetrated cultural walls with ease, and has strengthened and tightened connections between cultural groups. There is not even one culture that has not been influenced by another. Besides that, the advancement in science and technology in this age of computers and modern telecommunications has already planted a wide influence in patterns of thought and behavior including in those areas concerned with religion and theology.

For Kyai Tholchah, it is the spread of technology that is breaking down local identities. The above concerns resonate with Barber's observation that globalization threatens local identity (1995:82). However, Kyai Tholchah does not advocate avoiding technological development. In his oral presentation, he asked rhetorically, which parts of ASWAJA they wanted to preserve and promote. In the text of classical ASWAJA scholars are ideas that are clearly contradicted by science, like that of a geocentric universe. It cannot be these aspects of ASWAJA that are preserved but rather its value system. Another *kyai* in attendance agreed and said that concerns about the shape of the solar system are nonissues and that they must focus on values. Tholchah argued that it is possible to maintain traditional values and adopt technology.

Hence, the seminar also addressed possible strategies in the preservation and promotion of ASWAJA. Kyai Tholchah suggested three main strategies. His first suggestion was for intellectual reflection on the doctrines of ASWAJA. Some of the issues he suggested as needing reflection are considered closed issues by other *kyai*. For example, Kyai Tholchah asks

> May Sunni doctrine which is based on interpretation be reexamined or is it already unalterable dogma?

For many *kyai*, the "gate of interpretation" (*pintu ijtihad*) is already closed and so any interpretation of doctrine must be done within the framework of established religious scholarship.

Kyai Tholchah's second suggestion was for the preservation of traditional religious practices, such as *tahlilan* (prayers for the dead) and *selawatan* (short invocations), and the traditional values that are the soul of the *pesantren* world (1995). His third suggestion was for the development of religious infrastructure including more *pesantren*, mosques, religious publications, and public lectures. He argues that such educational activities are the key to the preservation and promotion of ASWAJA.

Kyai Djauhari agreed that education was the key and specified three educational programs: to introduce ASWAJA doctrine to the general population, to prepare quality instructors and motivators to lead the general population in practicing ASWAJA, and to produce books, magazines, films, and audio cassettes that teach about ASWAJA. He also stressed a need for greater Islamic brotherhood and solidarity. All of these activities, especially the first two are being done through *pesantren*.

Relationship to the Middle East

Islam was born in the Middle East. Major centers of learning have been in the Middle East, even when sometimes the leading teachers were Indonesian. The cities of Mecca and Medina remain the ritual centers of the Islamic world. Therefore *kyai* and their followers wrestle with what should be the relationship between Indonesian Islam in general, and the *pesantren* world in particular, to the Middle East and to Arabic Islam. As already mentioned, *pesantren* people are distrustful of the Wahabi movement in particular because of its de-emphasis on classical learning and texts. There are a number of other ways in which *pesantren* people carefully negotiate their relationship with Arabic Islam.

During the past decade or so, the question of what is the proper greeting to be used in Indonesia has been a matter of debate. Some have argued that Muslims should use the Arabic *Assalamu'alaikum* and hence advocate Indonesians to embrace a pan-Islamic identity and to follow the perceived practices of other Islamic lands. However, others including Abdurrahman Wahid, have argued that the Indonesian forms are sufficient and in so doing, they are declaring that the Indonesian Islamic community should maintain its Indonesian distinctiveness. In regard to greetings at someone's door, the other options include calling out the Indonesian "*Permisi*" or the Javanese "*Nyuwun Sewu*" both somewhat meaning, "Excuse me."

Some young Indonesian Muslims have taken to posting door stickers to encourage the use of the Arabic greeting. All of these stickers include imagery to further define the nature of Islamic practice. The most obvious of these shows a young girl wearing the *jilbab*, a head covering advocated by some Muslim leaders. It is telling that not all Javanese Muslim women wear this sort of head covering and so this sticker makes two claims as to the nature of proper piety (i.e., proper greetings and proper dress). Other images try to visually associate this Islamic practice with the Middle East. Some use rather silly looking camels, which are not specifically Islamic but are clean animals. Camels are not native to Indonesia and so are representative of the Middle East. However, the most interesting of these stickers calling for the Arabic greeting, depicts a cartoonish mummy and a pyramid. Although these images are of an imagined pre-Islamic Egypt, they emblematically connect the encouraged practice with the Middle East. Another interesting dimension of these stickers is a lack of Arabic (other than the greeting); the language used on the stickers is predominantly English although a good number are in Indonesian.

Another area of debate is in attire, especially women's attire. In recent years, more Javanese women have started wearing the *jilbab* (a head covering that covers the hair, neck, and ears, leaving the chin and face exposed). Jabar Adlan, the rector of IAIN Surabaya and of IKAHA, the college at Tebu Ireng is of the opinion that *jilbab* is just fashion. He says that the Javanese *kurundung* (a cap covering just the hair) is more than enough.

Islamic law requires both men and women to cover their *aurat*, which can be defined as those areas that are provocative to the opposite sex. Jabar Adlan argued that the concept of *aurat* was culturally bound and how the need for women to cover their head is Arabic culture. For men, the *aurat* is from just above the navel to just below the knees and hence is all that must be covered for prayers. However, it is better to use long-sleeved shirts, long pants or a sarong (some would argue a sarong is better because it does not clearly define the form of the legs and buttocks), and a head covering. If one uses these in everyday life it is better still. As for a woman's *aurat*, this is in-part culturally defined. Clearly, a woman's *aurat* would minimially include the same areas as a man's but would also include her breasts. More than that is open for interpretation. In practice, the *aurat* for women generally extends down to the ankle and to the wrists. However, Adlan argued, that because Javanese men are not aroused by hair, neck, or ears, the *jilbab* is not necessary in Java. On the other hand, he argued, Arabian men are so easily aroused that not only must women be completely covered, but also they are not allowed to perform public Qu'ran recitations because even listening to the holy text, Arabian men would be aroused by the sound of a women's voice. This evaluation of the moral status of Arabian men clearly established a hierarchy between Arab and Javanese men and favors Javanese practices over Arabic ones. It should be noted that not even all *nyai*, female *kyai*, wear the *jilbab* and not all *pesantren* require their female students to do so.

Another area in which the relationship with the Arabic world is negotiated is in language instruction. Traditional *pesantren* emphasize learning Arabic. Recently, *pesantren* have added instruction in English. For example, as one enters the Al-Hikam compound, there is a sign that requests all visitors to speak either English or Arabic. The hope is to create an environment in which students can learn these important languages. The model they used is Gontor (see Castles 1966; Steenbrink 1974), which has a strong "English or Arabic only" policy in which a network of spies, enforcement officers, and student courts prevent the use of Indonesian, Javanese, and other regional dialects. The leadership at Al-Hikam recognizes that they are still far from this goal.

The focus on English and Arabic instruction is one way in which the leadership of Al-Hikam is constructing modernity and tradition. English is the language of modernity and globalization and without it one cannot escape a peripheral position in the world economic and political order. Arabic is the language of a worldwide confessional community and is hence seen as a way of connecting Java, and more widely Indonesia, to the rest of the Islamic world.

The students spend two hours a week studying English in the *pesantren*'s modern language laboratory. The English program is supplemented by the presence of an English education major from a local teacher's training college who was recruited by Hasyim Muzadi to live at Al-Hikam so that the students would be able to practice their English. English is imagined as a key to international diplomacy and commerce; educating students in both the English language and Islam will create Indonesian diplomats and international businessmen who are grounded in traditional *pesantren* morality. In general, English instruction at the university level is either not offered or is considered an ineffective way of learning English. Throughout Indonesia there are hosts of small English language programs that are seen as the most effective way of learning the language. Al-Hikam students evaluated their English program as a viable alternative to the private programs.

The teaching of Arabic includes basic grammar, pronunciation, and calligraphy and is reinforced through much of the religious curriculum. Arabic is imagined as a link to traditional Islam through the ability to read the Qur'an, hadith, and the classical texts. It is also imagined as a link to the broader Islamic community and pan-Islamism.

Regarding the significance placed on Arabic by the Al-Hikam students, I recorded in my notes,

> I brought over some Nally's Cheese Balls I had bought at Hero's. The first question was whether this was an American snack. After being told that it was, the next question was whether it was *halal* [allowed]. I pointed out that one of the languages in which the ingredients were printed was Arabic and they concluded that it then must be *halal*. I suggested that they just read the Arabic ingredients list and that they would know. One fellow offered that an *ustadh* would be able to do that but that they would not.

Arabic is made synonymous with Islam: Arabic labeling means that it is allowable. Although given the existence of Arabic speaking Christians, it is logically possible to have a snack with Arabic labeling that is, in fact, not halal.

The general practicality *pesantren* people take toward Arab customs is illustrated in the following anecdote. When using my hands to share a common plate of food during a *manakib*, I used my thumb, index, and middle fingers to pick up my food. My friends asked me why I did not use all my fingers. I explained that an Arab friend had showed me to eat that way. I then suggested that perhaps this was also the way the Prophet ate. They respond that eating with three fingers would get one a lot of *pahala* (merit) because it imitates the Prophet, but using all your fingers would get one a lot of food.

The jury is still out about the question of cultural accommodation in Javanese Islam. Traditionally, the *pesantren* community in particular has been most willing to work with Javanese culture in spreading Islam. However, there are a greater number of *pesantren* people who are favoring Arabic dress, including robes, rather than the sarong for men. Political trends, including the U.S. invasions of Muslim countries have moved many Indonesians to identify more rather than less with their Middle Eastern bretheren. For now, it still seems that the general trend in the *pesantren* world is that Muslim practice should still look and feel Indonesian.

Negotiating a Modern/National Identity

In dealing with the Suharto government several questions were asked. What sort of relationship does the *pesantren* world have with the government? What sort of relationship should it have? What accommodations have been made between the government and the Indonesian Islamic community? How should *pesantren* people try to shape their collective identity as Indonesian, and their Indonesian identity as Islamic?

The Pesantren *World and the State*

In regard to the kind of relationship *kyai* see between themselves and the government, one *kyai* said to me,

> *Kyai* should not pay homage to the government, the government should pay homage to *kyai*.

In other words, the religious authority of *kyai* and *ulama* is greater than the authority of the state. In part, this is based on the notion that *kyai* are inheritors of the Wali Songo (the nine saints who brought Islam to Indonesia), who were instrumental in establishing all of the Islamic kingdoms in the Archipelago, a claim I heard repeatedly and one which Ann Kumar supports (1985:1–2). Hence, the legitimacy of the state should come from *kyai*.

Another aspect of the relationship between *pesantren* people and the Suharto regime is seen in their self-defined role as defenders of the people. In Indonesia, a gap has been constructed between "the government" and "the people." Government officials above the levels of the village and neighborhood are appointed from above and are often not even from the community or region to which they are assigned. Tsing argues that this gap is part of a colonial heritage. She states,

> The Dutch, for example, instituted the codes that now differentiate between "national" and "customary" [also religious] law; those rural people associated with "custom" are conceptually segregated from national administrations. . . . (Tsing 1993:25)

Kyai see themselves filling the gap between "the government" and "the people." In this regard, Yusuf Hasyim has said,

> The *pesantren*'s great potential for developing the lower classes has not only created opportunities for promoting rural communities but has also strengthened *pesantrens*' positions as a [*sic*] self-reliant, non-governmental organization. Morally, *pesantren*s belong to the public, and simultaneously become a model for social, political, religious, and ethical decisions. (1987:13)

During 1994–1995, perceived government favoritism toward Muhammadiyah led some *pesantren* people to feel some distance from the Suharto regime. This favoritism was seen in the declaration made by Suharto at the 1995 National Congress of Muhammadiyah, "*Saya Bibit Muhammadiyah*" (I am the seed of Muhammadiyah). Although in the past similar statements were made about NU, during my field season, NU was experiencing a rocky relationship with the government. Abdurrahman Wahid, its general chairman, was forbidden by government functionaries to speak at various local settings. Further, President Suharto had yet to acknowledge his reelection. This general condition inspired the editors of *Aula*, the magazine of NU East Java, to design the cover of the August 1995 edition (right after the Muhammadiyah National Congress) with the image of President Suharto in *ihram* (*hajj* dress) in front of the Muhammadiyah flag and the heading, "*Berbahagialah* Muhammadiyah" (Rejoice Muhammadiyah). Further, according to officials of *Lembaga Ma'arif*, NU's organization for school-based education, the government-approved religious curriculum advocates the Muhammadiyah interpretation of Islam. Three particular points of ritual practice were highlighted when I spoke with them: (1) in the performance of *tarewah* (special prayers in Ramadan after the final evening prayers), the curriculum teaches that only eight *rakyat* (prayer cycles) should be performed, whereas the NU (Shafi'i) practice is twenty-one *rakyats*; (2) that performing communal ritual meals (*slametan*) for the dead is not allowed, NU teaching allows *slametan*; and (3) that communal worship on high holy days (*Salat Eid*) should be in an open field, NU advocates the use of mosques. In response to this perceived bias in the government's religious curriculum, *Lembaga Ma'arif* designs curriculum guidelines for teaching around these problem areas.

Debates within the Islamic community regarding the relationship with the government began shortly after the establishment of the Republic of Indonesia, when there was great debate about whether or not Muslims could accept a secular government. The first draft of the Preamble of 1945 Constitution, known as the Jakarta Charter, required all Muslims to follow Islamic law (*shariah*) (Noer 1978:12). The "seven words" that granted such a requirement were deleted by the panel of nine writers, eight of whom were Muslims, including Wahid Hasyim of Tebu Ireng (Dhofier 1980b:163). This action established the Republic of Indonesia as a secular state (Abdullah 1996:65). This was contrary to the hopes of many *ulama* for the creation of a Indonesian Islamic State (*Negara Islam Indonesia*) when the Republic was born (Boland 1971:15–34). Hiroko Horikoshi states that such hopes were based on the fact that 90 percent of the population was Muslim and that Islam offered an alternative model of government to the colonial pattern (1975:60).

A further symbolic blow to the Muslim community was a change in vocabulary for the title of the preamble; the Arabic word for preamble, *Muqadimah* was rejected in favor of the Indonesian *pembukaan*. Deliar Noer suggests that this move communicated to the Islamic community that things "regarded as Islamic were contrary to national aspirations" (1978:12).

The resentments growing out of the above chain of events led to the *Darul Islam* (State of Islam) movement that was strongest in West Java[3] (Horikoshi 1975), but also had a strong presence in Central Java, South Kalimantan, South Sulawaesi, and Aceh (Nasution et al. 1992:197). Such movements were not finally put to rest until 1962 (Horikoshi 1975). NU itself was committed to establishing *shariah* as the basis of government, but made a compromise with the nationalists in order to establish the Republic of Indonesia. Wahid Hasyim negotiated a compromise that dropped the requirement that all Muslims follow *shariah*. In exchange, the first principle of the Panca Sila[4] (Five Principles), was changed from Belief in God (*ketuhanan*) to Belief in the One and Only God (*ketuhanan yang maha esa*) (Feillard 1997:132). Today, all but the most ardent have abandoned the goal of an Islamic state (*Negeri Islam*) and have chosen instead to focus on strengthening the place of Islam among the populace *(Masyarakat Islam)*. Individual *kyai* have specific and varying ideas about how to accomplish this goal.

The Panca Sila is the ideological basis of the Indonesian state. The first principle is monotheism (*Ketuhanan Yang Maha Esa*), but monotheism has been defined in such a way that the five recognized religions (Catholicism, Protestantism, Islam, Buddhism, and Hinduism) are all monotheistic. This definition was disappointing to many Muslims who thought that Islam alone recognized the sole unity of God (Noer 1978:13fn). This disappointment followed closely on perceived victories: the rearrangement of the principles so that the issue of belief became the first principle instead of the fourth and the changing of the principle to just monotheism rather than a broader "belief in God" (including polytheism) (Madjid 1996:89).

In 1955, Indonesia sought to redefine its constitution and philosophical basis. The Constituent Assembly was deeply and irreconcilably divided between the nationalists, the communists, and the Muslims. When the assembly came to deadlock, President Sukarno, with the backing of the military, disbanded the Assembly, returned the nation to the 1945 constitution, and declared that the Panca Sila was the permanent philosophical basis for the nation. In a gesture of accommodation to the Muslim's wish for a state based on Islamic values, "Sukarno declared that the 1945 Constitution and the Panca Sila should be seen as the historical continuation of the Jakarta Charter" (Madjid 1996:90). Nurcholish Madjid states that the Islamic community did not receive this decree with much enthusiasm (1996:90).

Following an allegedly communist coup attempt and a great bloodletting in 1965–1966, Panca Sila became more firmly entrenched when the emergent President Suharto established it as the "only ideology to guide the Indonesian people at the national, political and societal levels" (Madjid 1996:91). Beginning in 1978, the Republic of Indonesia began an intensive program to train its citizens in Panca Sila ideology. By 1984, all social organizations had to accept the Panca Sila as their sole ideological basis (Weatherbee 1985:190), hence moving the jurisdiction of Panca Sila to more specialized aspects of social life. Weatherbee states,

> It was seen as a major attack of any legitimate Islamic voice in critical matters of human affairs other than the narrowly religious business of the mosque and *fiqih* (Islamic law). (1985:190)

Elsewhere, I have discussed how non-*ulama* Muslim intellectuals have accepted the government ideology of Panca Sila and government goals in general. But they have, through their manipulation of metaphors, made the statement that Islam is best able to meet these goals and realize this ideology (Lukens-Bull 1996). Similarly, *kyai*, through their educational efforts, are claiming to be more capable to move the Indonesian people into the twenty-first century.

Several political clashes with the government in the 1970s (over civil marriage for Muslims, restrictions on polygamy, attempts to recognize mystical groups as equal to officially recognized religions, and others) lead to a tense relationship. This tension eventually led to NU withdrawal from practical politics and the *Kembali ke Khittah* 26 (Return to the Spirit of 1926) movement (Feillard 1997:135).

Nahdlatul Ulama's return to moderation was indicative of a public accommodation between Muslims and the Indonesian state. In this accommodation, the government conceded that: the government would encourage worship and ceremonial practices; the state would support and administer education, pilgrimage, and other communal matters. In return, the leaders of the Islamic community conceded that religious organizations would not be involved in electoral politics and that they would support the government's national development policies through their various teaching and preaching activities (Federspiel 1996:193–194). However, this was merely the public face of the relationship between the government and the Muslim communities. As Hefner observes, "Many [were] troubled by their accommodation with the president and recognize that theirs is an alliance of unequals in the extreme" (1997:111).

In NU, there were many *kyai* who were not happy with the withdrawal from practical politics and using the Panca Sila as the basis (*asas tunggal*) of their organization. Although they publicly accepted this, they were biding their time and were ready to move back into politics when Suharto fell (van Bruinessen 2002).

Likewise, the Suharto regime sought to manipulate conservative Islam against Muslim moderates and political reformers like Abdurrahman Wahid. One example of this is ICMI (The Association of Muslim Intellectuals) that drew mostly Modernists. Although the media accounts of the founding of ICMI attribute it to the efforts of five college students at Universitas Brawijaya in Malang, Hefner demonstrates how several New Order figures including Suharto himself were involved in back room negotiations to create this organization (Hefner 1997:94–99).

Since ICMI was backed by Suharto, several *kyai* saw it as a potential pawn. In September 1995, key figures in both ICMI and the army supported Yusuf Hasyim's continued efforts to work against his nephew Abdurrahman Wahid (Barton 2002:211). One night in late 1995, a small group of East Java *kyai* gathered to discuss how to infiltrate and subvert ICMI. Whether they did this in direct response to the anti-Abdurrahman efforts or simply because they saw ICMI a general threat is unclear. What is clear is that they were sufficiently concerned about the inherent political nature of their actions that although they invited me to observe, they foreswore me to never report their names or the details of the meeting. They did, however, permit me to discuss it in the vaguest of fashions.

The *pesantren* community also sought to manipulate this public accommodation. They accepted the idea that rather than use politics, they had to find other ways to attain the religious goal of creating an Islamic society (*masyrakat Islam*). Of course, a main way to do this was through education. Islamic education was a chief means for *kyai* to outwardly support the national development policies while striving to firmly establish Islamic values as the foundation for public life in Indonesia. One school official put it most subversively, "We accept the government program, but train the students with our values, so that we can place enemies of the government within its very ranks."

National Identity and the Pesantren *World*

Indonesian identity is being constructed in *pesantren* education. There are several ways this is being done, including such obvious nation building techniques as conducting government education in the national language, Indonesian. However, less obvious is the use of Indonesian in *ngaji* education. Traditionally, the meaning (*makna*) of classical texts was given in a regional language (Javanese or Madurese). However, this is changing. In part this is because student populations now include greater numbers of students from outside the region, and hence, not all students know Javanese or Madurese. In order to understand the Arabic texts, they must learn yet another language. At some *pesantren* then, the beginners hear the Arabic text, the Javanese *makna*, and an Indonesian explanation. Advanced students are still expected to learn Javanese in order to study the texts. However, *musyawarah* amongst them are often conducted in Indonesian rather than Javanese. Other *pesantren* have opted for giving the *makna* directly in Indonesian and have trained their students to write Indonesian in Arabic script. The use of Indonesian rather than a regional language moves *pesantren* education from being a Javanese (or Madurese) experience to being an Indonesian one. Hence, any Indonesian may receive this education and be inducted into the *pesantren* community.

Discussion

This chapter has explored various elements in the identity debates in which *pesantren* people are engaged; taken with the previous chapter, we have a summary of all the elements of these debates. *Pesantren* are educational institutions. Once they were the only form of education available and exclusively religious in nature; now debates about the curriculum are rampant as some *pesantren* are adopting secular education. *Kyai* are educators through *pesantren*, as well as, through their lifestyle examples and preaching. As examples for Islamic living, the identity of *kyai* shapes the identity of the community. *Pesantren* people (*kyai*, their *santri*, and other followers) are part of broader religious communities. NU is a national organization that ties *pesantren* and *kyai* together; although this organization had its origins in East Java it now connects Classicalist Muslims from all over Indonesia. *Ahlus sunna wal jamaah* is an identity marker that ties the *pesantren* world to the worldwide Islamic community and simultaneously makes a claim of religious authenticity; not only are *pesantren* people Muslim, but they are also part of the "real Muslims." Finally, the identity of *pesantren* people is not limited to religion; they are Indonesian citizens and engage that identity wholeheartedly, seeking to Islamize it as best as possible. The next chapter turns to an analysis of the processes of identity contraction and the (re)invention of modernity and tradition as it manifests itself in terms of the nature of *kyai* leadership in modern Indonesia.

Chapter 5

You Are Who You Follow: Negotiating Leadership and Community Identity

I was sitting in a coffee shop with a group of men talking about their small town. When the conversation turned to the local *pesantren* and the men who ran it, they fell quiet. Only after I assured them that I was not working for any *kyai* and that no one save me would ever see my notes, did they slowly come to tell me that they did not like these three *kyai*. It then dawned on me that unlike how most Javanese characterize them, at least in public, *kyai* were not necessarily universally revered. Even after my assurances that I would take steps to protect their identities, they spoke in low voices and glanced about the room furtively. Their concerns suggest that criticisms of *kyai* are best relegated to whispered conversations; to drag them into the public sphere would create difficulties no one wishes to even consider.

As it turns out, the Javanese have a complex set of criteria by which *kyai* are evaluated. As "models of behavior," *kyai* are scrutinized and evaluated, although this aspect is not always part of the public discourse. Because of the nature of *kyai* leadership, criticism sometimes must be part of hidden transcripts (cf. Scott 1990). Although the debates may not always be open, the criteria as well as whether a particular *kyai* meets a particular dimension of them is hotly contested. Of particular interest is the relationship between "*kyainess*" (*kekyaian*), that is, the categorical identity of *kyai*, the identity of individual *kyai*, and the collective identity of the Classicalist Islamic community in Java and throughout Indonesia. *Kyai* can shape the identity of the community through their personal identity construction, as models of behavior.

Thus far, we have explored the nature of tradition and modernity in the Classicalist Muslim community in East Java. I have explored how they are (re)invented in education as well as how the (re)invention of modernity and tradition ties into multiple layers of community identity. Now, I shall investigate how these processes have affected *kyai* leadership. To do this, I will first explore how this community evaluates *kyai* and their claims to authority. This is particularly important to consider because modern education as well as participation in the state suggests alternative models of authority. I then examine how all this played out when a *kyai* found himself elected president of Indonesia.

The debates about the nature of *kyainess* came to forefront when Abdurrahman Wahid served as President from October 1999 to July 2001. As an emerging democracy and a Muslim majority society (>80 percent), there is considerable social discourse and debate in Indonesia about the nature of leadership, politics, and nationhood. It can be said that Gus Dur was elected as much for the culture category he occupies as for his personal qualifications. His being a *kyai* continues to be a factor in how many people continue to evaluate his presidency. However, people's opinions are very much split. His followers also invoke these traditions and some have gone as far as to call him a living saint. Some accept that he is *kyai* but argue that the trappings of *kyai* leadership do not belong in the office of the presidency. Others use the standards (*fikih* texts, *kitab kuning*) used by *kyai* to judge him even though they reject these texts as sources of religious knowledge in general. Others reject him as a *kyai* and reject the whole notion of *kyainess*. Further they claim that he is just using the symbols of *kyai* leadership to legitimize his rule in the eyes of the masses in spite of *kyainess* not being meaningful to him personally.

I became interested in the Gus Dur presidency because I had known him before his involvement in electoral politics. I knew him as the eccentric *kyai* who was well respected in the *pesantren* community for number of reasons including his oratory skills, his sense of humor, his genealogy, his religious acumen, his spirituality, and his opposition to Suharto. I found myself in a desirable position to evaluate the interaction between traditional and modern leadership models. In 1995, I had collected extensive data about the process and criteria by which Javanese Muslims evaluate *kyai*.

Gus Dur is Judged by *Kyai*

Abdurrahman's presidency was not the first time he had come under fire for his odd behaviors and extraordinary opinions. Toward the end of Gus Dur's first term as the general chair of Nahdlatul Ulama (NU) (1984–1989), there were concerns about a number of his actions and expressed opinions. A meeting was held in Cirebon, West Java on

March 8–9, 1989. Over 200 *kyai* attended to express their concerns. The concerns included: (1) his apparent opinion that Islam is not sufficient to deal with societal affairs; (2) tolerance for Islamic heresies; (3) his ideas about localizing Islamic practice; and (4) his becoming part of the Jakarta Arts Concil. Gus Dur responded to these. Out of this meeting came a short book *Gus Dur Diadili Kyai-Kyai*, or *Gus Dur is Judged by Kyai* (Hamzah and Anam 1989), which was reprinted in 1999 on the occasion of Abdurrahman's election as president. In the words of one of the editors, it could provide the nation with insight to its new president. The proceedings of the original March 1989 meeting only fill about 21 pages. The rest of this volume (about 108 pages) is drawn from Abdurrahman's writings, which have appeared in a number of newspapers and magazines.

Before responding to the specific concerns of the assembled *kyai*, Gus Dur gave a long explanation about *khittah NU* (the spirit of NU) and the decision of NU to pull out of party politics and return to its roots as a socioreligious organization focused on education, proselytization, and social justice. Among the reasons given here as to why NU pulled out of politics was the level of control the Suharto regime exercised over the two approved "opposition" parties.

He suggested that NU had to prepare its membership and followership to face the future that included both religious and nonreligious issues. He suggested that an important issue was the place of *shariah* in Indonesian government and society. He separated *shariah* into that which must be put into law and that which would be sufficient to make part of the ethics (*akhlak*) of the general public. He argued that NU was required to be prepared in the economic arena and that the future would require a high degree of economic growth based on a high level of export trade and meeting the national consumer needs as cheaply as possible. He argued that this economic goal requires objectivity, justice, and the rule of law. He emphasized the need for Indonesia to abandon nepotism.

As to Islam's role in society, he argued that the Islamic community needs to be interested in social justice and he does not understand why they are not. He basically argues that development should be understood as the greater jihad.

As far as the place of Islam in government is concerned, government positions can be held by non-Muslims, but the Muslim community should try hard to sell themselves so that all government positions are filled by Muslims. He argued that NU should be sure that others are not seen as flexible whereas NU is seen as rigid.

In regard to his opinions about Mutazilah and Shia, he makes the distinction between understanding the position of a particular group

and accepting it. As to Mutazilah he only admired their principle of
justice, which is only one of five. As far as his position on Shia, he had
been misunderstood, but has simply stated that NU loves the family of
the Prophet and Shia does too, even though there are differences.

In regard to localization, he argued that such initiatives are the
heritage of the Wali Songo and that NU must take this up. Starting in
the late 1980s, Abdurrahman became increasingly involved with various dimensions of Javanese spirituality, sometimes pushing the enve-
lope even for a Classicalist Muslim, and clearly doing things decried by
Reformists. He made visits to the graves of many Javanese Islamic
saints and even to the graves of people whose legitimacy as Islamic
figures are questioned or outright refuted (van Bruinessen 2002).

On becoming the head of the Jakarta Arts Council, he reminded
everyone that he accepted the position before becoming head of NU.
He could not quit mid-term, but his term ended six months into his
first term as head of NU. He said that he took the position originally
so that the moral and religious dimensions of art could be assessed.
He argued that artists were starting to understand Islam, and that it
would be a pity if no one in the *pesantren* milieu paid any attention to
these efforts. He also stated that the Minister of Information, Harmoko,
was glad to have him aboard to prevent the possibility of a film win-
ning an award but showing immoral activities (i.e., wife stealing). To
complete his obligations he had to serve on an arts jury three times,
he did this and quit. He obeyed the wishes of KH Asad Symasul Arifin
to do so, even though many on the arts council were disappointed.

In regards to opening a Jesus Christ Poem Night, he said that what
is forbidden is participating in another religion's *ibadah* (rituals) and
this was not that. Further, Jesus Christ (*Yesus Kristus*) is not a name
that is necessary linked to a particular faith (*akidah*). He argued that
addressing this figure as Jesus Christ, does not make one a Christian;
in Arabic he is called *Isa al-Masahih*, Jesus the Messiah, which means
the same as Jesus Christ.

The most important observation from this discussion is that even
to his peers, Abdurrahman can be an enigma. However, this very
enigmatic nature proves to some people that he is a living saint (*wali*).
Apparently, the concerns were met sufficiently; Gus Dur was reelected
in 1989 and again in 1994 and served as the general chairman of NU
until he assumed the presidency of Indonesia in October 1999.

The Cultural Category of *Kyai*

The debates about Gus Dur both in 1989 and ten years later when he
was president are grounded in general cultural understandings about

kyai. The term *"kyai"* in Javanese is used in several ways. However, in all these ways it is used to indicate something or someone that has supernormal (*ghohib*) qualities. One *kyai* said that the etymology of the term is from the Javanese *"iki wai"* (which he translated as "this one to be chosen") indicates that *"kyai"* are special because they are chosen by Allah. However the term *"kyai"* can be applied to nonhumans, indeed several items of royal Javanese heirlooms (*pusaka*) are called *kyai*, including *kris* (Javanese long daggers) and horse carriages used by the royal family. Further the term *"Ki"* is said to be cognitively and semantically related and although it is occasionally used for Islamic holy men (cf. Muhaimin 1995:123), it is usually reserved for the puppet master (*dalang*) of the Javanese shadow theater (*wayang*).

The term *"kyai"* can be applied either to a category or an individual in that category (see figure 5.1). It is also used as a term of address

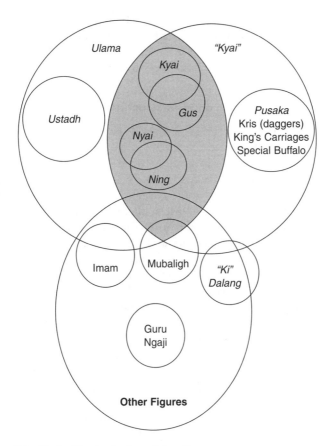

Figure 5.1 Kinds of *kyai* and other religious figures

for a specific person in the category *kyai*. Not all individuals who belong to the category are called by the term of address, "*kyai*." Sons of *kyai* are called "*gus*," in part to recognize their status as members of the category *kyai*, as sons of *kyai*, and in part in recognition that they have not earned the right to the title in their own right yet. Some *kyai*, Gus Dur for one, are affectionately called *gus* to recognize their heritage as sons of *kyai* even though they sometimes are addressed by the title *kyai*. The wives and daughters of *kyai* also belong to the category; wives are addressed by the term "*nyai*"[1] and daughters by the term "*ning*." Many *nyai* are renowned for their mastery of texts and their ability to teach (mostly, if not exclusively women) and their special skills in healing. Hence *nyai* belong to the category of *kyai*.

Besides *kyai*, there are other, lower ranking Islamic leaders in Java. First, is the *ustadh* who is typically a senior *santri* but is almost always under the authority of a *kyai*. Second, is the Imam Masjid, or Mosque Leader, who leads worship and who may be responsible for the physical plant of the mosque and for distributing the alms collected there, but is not otherwise a significant community leader. Third, are the men and women who teach basic Qur'an recitation.

Contesting and Constructing *Kyainess*

The title *kyai* is not conferred, it is used. However, it is a breach of etiquette to use it for oneself, at least until one's position as *kyai* is well established. The title is said to come from the people, that is, they recognize someone as having the necessary qualifications and start addressing him as *Kyai* so-and-so. Since the authority of a *kyai* is allegedly bestowed by his followers, there is a dialectic process of claim, evaluation, and counterclaim. This process generally focuses on specific *kyai* and their behaviors and works to a general discourse about what *kyai* are and what they should be, which includes five components: institutional minimum, knowledge, spiritual power, genealogy (both spiritual and biological), and morality. However, morality is paramount as knowledge and power are linked to it. Knowledge gives rise to morality or else it is false knowledge. Spiritual power comes from *karoma* (closeness with God) that arises from religiosity and morality. And although the sons of *kyai* are expected to have higher moral standards, it is widely acknowledged that this is not always the case. Hence, not all sons of *kyai* achieve the knowledge and *barakah* needed to be considered *kyai* themselves.

There are certain moral and intellectual requirements for being a *kyai* and certain ways in which a *kyai*'s morality and learning are said to manifest themselves. The degree to which there is freedom to engage in discourse about these requirements may be a measure of whether we can consider *kyai* members of a class that have risen to defend the interests of that class or whether they are best considered an elite class. In part, the claim that the title "*kyai*" is from the people is a form of reversal designed to invoke communitas and anti-structure and thereby maintain the structure (cf. Turner 1969).

Institutional Minimum

In defining the role and character of *kyai*, the leadership or ownership of a *pesantren* is critical (Dhofier 1999:31, 39). *Kyai* who do not have a *pesantren* are rare and those who do exist are either in the process of establishing one, for example the famous preacher Zainuddin MZ was in this process during my field season, or they have a *pesantren* attributed to them. Abdurrahman Wahid (Gus Dur) is an example of the second case and has *pesantren* attributed to him both literally and metaphorically. Literally, he is said to be the headmaster of either a small *pesantren* by his house in South Jakarta (which I could not confirm) or the headmaster of Tebu Ireng (which is incorrect). Metaphorically, NU was referred to as a *pesantren* with Gus Dur as its director.

Kyai are characterized by "*nilai lebih*" (personal communication, July 2001) (higher value) or a sense that they are more learned in religious matters. Kyai Badruddin stated that if one person teaches another religious subjects, then the first person is *nilai lebih* to the other person. Abdurrahman Mas'ud suggests that there are some famous *kyai* from Demak who do not have a *pesantren*, but do teach maintaining the relationship of *nilai lebih*. However, in general, having a *pondok* seems important in that it provides people to whom the *kyai* was more learned. This claim implies the axiom that a *kyai* without a *pesantren* is not really a *kyai*. This axiom's corollary (a *pesantren* without a *kyai* is not really a *pesantren*) is seen in the claim that Habibe's *pesantren* was doomed to fail.

The Powers of Kyai

Kyai are believed to possess *barakah* that often leads to reports of supernormal or magical power (Dhofier 1980a:53; Kartodirdjo 1966:163; Moertono 1968:80). *Barakah* comes to *kyai* in part because *ulama* are *penerus Nabi* (inheritors of the Prophet). Clifford Geertz discusses how the basic concept of *barakah* as blessing is linked to ideas of material prosperity, physical well-being, bodily satisfaction, completion, luck, plenitude, and magical power (1968:44). About *kyai*, Zamakhsyari Dhofier states,

> Most *kyai* also foster the public image that some *kyai* are extraordinary
> persons possessing the spiritual attributes of *karomah* (a man who is
> greatly loved by Allah) and so becoming sources of *barakah* (Allah's
> blessing) for their followers. (1980a:53)

Barakah is evidenced by a "personal presence, force of character,
moral vividness" (Geertz 1968:44). Because *barakah* can be transmitted
from its holder to others, a *kyai*'s or a *syehk*'s popularity and authority
is partially derived from his value as a source of *barakah* (Barth
1959:11; Dhofier 1980a:53; Eickelman 1976:6; Gilsenan 1982:86,
109). After death, the *barakah* can linger in the body and is transmitted to the area around the tomb (Geertz 1968:49). The tombs of men
(and sometimes women) possessing *barakah* form pilgrimage sites
throughout Java and the rest of the Islamic world.

Kyai are said to have special powers. A sick person can be healed by
the mere touch of a *kyai* or by being given a drink of water by a *kyai*.
Sometimes these illnesses are caused by *jin* (genies), spirits, and other
spiritual causes. *Kyai* also have the ability to kill people or stop trains
by just speaking the command. These abilities come from *barakah*
that comes from Allah (Barth 1959:11; Dhofier 1980a:53; Eickelman
1976:6; Gilsenan 1982:86, 109; Kartodirdjo 1966:163; Mansurnoor
1990:260, 324; Moertono 1968:80). Some *kyai* can walk through
the rain and not get wet. *Kyai* are often seen in two places at the
same time.

Once I observed Bu Nyai Masduqi (the wife of the then vice chair
of the *Syuriah* branch of NU for East Java) receiving guests, a young
woman, her mother, and her brother. As the family was getting ready
to leave, the Bu Nyai told them to drink the beverages they had been
served earlier in the visit. They had not drank any yet. As the young
woman tried to drink her beverage (a sweet fruit juice often served
there) she contorted her face in reaction to its "bitterness." The *nyai*
then told her to drink it all, as the *barakah* was in the bottom. When the
family left they were given a package of "medicine" complete with a
set of verbal directions. As the woman left, the *nyai* took her head,
rubbed it, and then blew on it. After the family left, the *nyai* left the
room. I asked her son, who was witness to this also, what had just
transpired. He said that the woman was sick with confusion (*bingung*)
and stress because she was possessed by *jin*.[2] They came to ask for *air
selawat*, which is any drink over which the *selawat* (history of the
Prophet) had been read. The drink was bitter to the woman because
of the possessing *jin*'s reaction to the "holy water." I have observed

other *kyai* affecting both physical and psychological healing by dispensing such beverages. Pak Fikri, a department chair at Universitas Muhammadiyah relates how when he first sent his daughter to Al-Fulan, she was unhappy and kept coming home. Kyai Badruddin gave her *air selawat* and now she would rather be in the *pondok* than at home. Such powers of healing and control of *jin* and other spirits are just part of the powers of *kyai* that emerge from their *barakah*.

It is possible for someone to imitate a *kyai*. This imitation is called *kyai-kyaian* ("toy" *kyai*) in which someone comes from outside (a certain village) and acts like a *kyai*. If he acts like a *kyai*, the people will call him *kyai*. However such impostors do not use Islamic knowledge (*ilmu*) but rather *sirih* (sorcery) or *jin*. These impostors invariably fail because they lack *barakah*. A related issue is the various *pesantren* established by non-*kyai*, which are said to always meet with failure. Informants told me that Muhammadiyah has attempted to establish *pesantren* so as to strengthen its rural, popular base. However, Amien Rais cannot successfully have a *pesantren* because he is not a *kyai*, he does not have *barakah*. The most famous example of a failed *pesantren* is that established by B.J. Habibie, then Minister of Technology. Informants said that this failed because Habibie went about it backwards; he built the grounds and then sought teachers and students. Further, there was no chance of success because he lacked *barakah*; he was not a *kyai*. Reduced levels of *barakah* or *karisma* (spiritual attractiveness) are sometimes attributed to sons and grandsons of *kyai*, who then argue to be "coasting" on their ancestor's reputations.

Knowledge of Kyai

In Java, as in many Islamic regions, certain families are seen as having acquired special religious knowledge (*ilmu*) that can be transmitted only through particular descent lines (Gibb and Bowen 1958; Green 1978; Horikoshi 1976:200; Keddie 1972; Mansurnoor 1990:237; Metcalf 1982). Further, a *kyai*'s intellectual connections are important to his position of respect. At the very least, *kyai* will acknowledge their educational history. Sometimes, intellectual genealogies (*silsila*) are traced back to the Prophet. Many display their *silsila* and often proclaim from whom they studied.

Related to this last point is the importance of family connections in establishing a particular person as a *kyai*. Certain families are known for their religious leadership. For example, Abdurrahman Wahid and several other prominent leaders in NU are the sons and grandsons of the key founders of the organization. This factor is often mentioned in evaluating their legitimacy.

In addition to having *ilmu*, *kyai* must be able to "*membawa masyrakat*," which literally means "carry the people." I was told that this means that he knows what people are like just by looking at them. For example, he could step into a dispute and immediately know the situation and set everything straight by dispensing his opinion. *Kyai* are thought to always speak the truth and give good advice. For example, a *kyai* can accurately predict a good place to dig a well or give successful advice about dealing with a misbehaving child. Because of this, *kyai* are often informal community leaders and opinion shapers, a fact noticed and utilized by some development agencies (Clement 1987:2).

A *kyai's ilmu* is also measured in part by their ability to speak well. This is important enough that I have encountered *kyai* who never speak publicly, but their *santri* feel it necessary to assert that they are skilled speakers. Religious knowledge (*ilmu*) is not limited to book knowledge (*pengetahuan*) but can include esoteric knowledge (*ilmu laduni*), including the ability to heal the sick, stop trains, and perform other miracles. A *kyai's* special abilities come from *barakah* (blessing from Allah), which rise from the *kyai* having a special relationship with God (*karoma*), which comes from superior morality. Such *ilmu* is thought of as having an existential nature. For example, Kyai Hamad of Al-Fulan is said to have stored his *ilmu* in the ocean once and the ocean calmed until there were no waves.

Abdurrahman Wahid is said to have a number of spiritual abilities. He is supposedly able to predict the future and predicted his presidential election. One story claims that he was in Pekalongan, a center for high quality batik, in early October 1999, prior to nomination as a presidential candidate. When asked what he was doing there, he is reported to have replied that he needed to buy batik for his new job as President of Indonesia. He has, on a number of occasions, used esoteric knowledge to find abandoned graves of saints and commanded local villagers to care for them in the future so that their village will once again be prosperous.

Morality of Kyai

Kyai are held to a higher standard of morality; if they indeed have greater knowledge and a special relationship with God, then they should be more moral. Imron Abu Amar in a book on popular Indonesian piety, delineates four levels of piety. The first level *Taqwal 'Am*, is the piety of the general Muslim believer. They do not commit *shirk* (associating others with Allah), but still occasionally commit major sins and often commit minor ones. The second level, *Taqwal 'Alim*, is the piety of those who are learned in Islam; that is the majority of *ulama* and *kyai*.

They only commit minor sins on occasion. They are still given to committing acts, which although not sins, are not favored by Allah (such as smoking). The third level, *Taqwal Arif*, is the piety of saints, called Wali (Friends of Allah) in Islam. Some *ulama* and *kyai* have achieved this level of piety. They do not commit sins and are quick to please God. The final level, *Taqwal Anbiya*, is the piety of the prophets. They neither fall into sin of any degree nor do they commit acts that are unfavored by Allah (1989:25–26).

About Wali, which includes some *kyai*, Syehk Yusuf bin Sulaiman states in the *kitab, Jami'u Karmatil-Auliya*,

> A Wali, or Friend of Allah, is someone who is very close with Him and because of his devotion Allah give him powers as well as *karomah* and protection. (As cited in Abu Amar 1989:23, translated from Indonesian)

The protection mentioned above includes protection from sin and immorality. KH Hasyim Asy'ari, in his *kitab Ad Durarul-Muntantsirah*, states that a Wali will be protected from

> committing sins, big or small. Being led into sin by his desires. If he should happen to sin, he will quickly return to Allah and repent. (As cited in Abu Amar 1989:23)

The above passages suggest that *kyai*, particularly major *kyai* should be above reproach. Hence, *kyai* are evaluated by their morality. Criticizing a *kyai*'s morality, is in effect questioning his legitimacy and his relationship to Allah.

The moral standards for *kyai* and how they are applied are contested. For example, most in the *pesantren* world would agree that greed, grudge holding, and revenge are immoral. However, there will be disparity in evaluating a particular *kyai* or a particular action. What one person might call grudge holding, inhospitable behavior, or general unfriendliness, another might see as righteous indignation. Other issues such as eating at a food stall in one's own neighborhood might not be included in all definitions of morality, but are generally considered to demonstrate a lack of refinement, in this case a lack of appreciation and love for one's wife. Political orientation as a measure of morality is, of course, the aspect most open to debate. There are *kyai* who were roundly criticized and despised, at least by some, for their affiliation with Golkar (the governing party). There are other *kyai* who were hailed for precisely such an affiliation.

Kyai who do not measure up are criticized and subject to gossip. Slamet, a tailor and entrepreneur who had previously studied in a *pesantren* but now was lax in his observance of the Five Pillars, told me the story of a *kyai* who held a grudge even though he should have been above that. The *kyai* in question was a friend of his from his *pesantren* days who had married the daughter of a famous NU figure, and hence was now himself a major figure. Once when this *kyai* was visiting family in Slamet's hometown, he encountered Slamet and asked him how he was doing. Slamet replied that he was not feeling well and excused himself and did not engage his friend, the *kyai*, in conversation. Slamet says that even many years later, this *kyai* will not greet him but will greet someone standing next to him; a great affront in Javanese and Islamic etiquette. Slamet complained that as the representative of the Prophet, a *kyai* should be able to control irrational emotions.

The importance of *kyai* being good examples of Muslim practice both for their students and the general public was expressed in a saying that I heard several times: "If the teacher pees standing, the students will pee running" (*Kalau guru kencing berdiri, muridnya kencing berlari*).[3] *Kyai* who are busy with many activities outside their *pesantren* worry that they are not able to provide for their students and neighbors the necessary living example they need in order to lead pious lives themselves.

On another occasion, Slamet took me over to Edy and Tuti's house. Tuti is an alumnus of Al-Fulan[4] who graduated in 1976. Edy and Tuti were of the opinion that Al-Fulan no longer had much influence on the populace around it because the current *kyai* failed to be a good example (*contoh*). Kyai Fulan, the founder, was *contoh* but his sons are not. When Fulan died, the *barakah* in that family was severely diminished. Fulan's sons do not socialize with the general population, neither do they do *da'wa* (propagation of the faith). Edy said that the problem with "*kyai turunan*" (*kyai* who are so only because their fathers were *kyai*) is that they did not come up from the masses, so they are out of touch. When Fulan died, Edy and Tuti ended their affiliation with Al-Fulan, despite being neighbors.

Being sociable and accessible is indeed an important measure of a *kyai*'s character. Gus Asyrof, the grandson of Kyai Fulan, said that his maternal grandfather would go to anyone's funeral, *slametan*, or other invitation (*undangan*); that it is the character of *kyai tasawuf* (Sufi *kyai*) to not pay attention to the social position of people. Asyrof said that his grandfather was a simple man, but that his sons were not. He said that he is trying to follow his grandfather's example. He said that he has declined an invitation from a government official to attend a neighbor's funeral. Asyrof may indeed be meeting his goal, for the cynical Slamet approves of him alone among all of the *kyai* at Al-Fulan.

In the summer of 2000, I attended a memorial service for Kyai Fulan in which one of the guest speakers, a high-ranking *kyai*, worked into his speech a criticism about the sons along these lines. He declared that Kyai Fulan's sons are more knowledgeable then he was, but that no one, including his own sons, surpassed him in the matter of moral behavior. He related how Fulan frequently visited him (again showing the morality of sociability), but that his sons never have; their moral behavior is not the same. The audience erupted into laughter and my neighbor whispered conspiratorially that everyone knows this about them, thus echoing the complaints that I heard in 1995 from people who lived in the neighboring village.

Another point over which *kyai* might be evaluated is consumerism. Slamet once complained about the poor pay for teachers in schools affiliated with *pesantren* (much lower than public or other private schools). His brother teaches physical education at an Al-Fulan school and in mid-1995, he was seriously considering going to Saudi Arabia to become a chauffeur. Slamet argued that it was wrong for teachers to be so poorly paid and for the *kyai* at Al-Fulan to own several cars each. When I suggested that for some *kyai* at least, cars were a necessity so that they may travel to all their speaking engagements, he pointed out that these *kyai* have few speaking engagements. Further, he related how a friend drove one of the Al-Fulan *kyai* to Surabaya to look for antiques. He angrily questioned where all the money goes: the money from the student's parents, as well as the money from in-house stores (which supply all the needs of the *santri*). Indeed, it is a violation of Islamic law (*shariah*) for a teacher to gain personal wealth from their position as teacher and leader (Dhofier 1999:153).

The Politics of Kyai

The involvement of *kyai* in party politics is another point over which individual *kyai* are evaluated. For those to whom a *kyai*'s politics are important it is more than a political question, it is a moral one. Supporting the "wrong" party, or "wrong" candidate is often seen as caused by a moral flaw.

During 1995, some criticized Abdurrahman Wahid for making public appearances with then head of PDI (Indonesian Democratic Party), Megawati Sukarnoputri. He denied any political connection; that they were simply friends and that their association should not in any way be taken as a sign of an allegiance between NU and PDI. Nonetheless, Gus Dur was criticized for his antagonistic relationship with the Suharto regime; some other *kyai* sought to have him removed from his position in NU because of it.

While Gus Dur received some criticism for being too antagonistic toward the then ruling party (Golkar) and too friendly toward opposition parties, other *kyai* were criticized for working with Golkar. I talked to a group of village men who live near the pseudonymous Al-Fulan about their opinion of it. Although somewhat reticent, they said that they disapproved of Al-Fulan and its current *kyai*, Hamid, because he supported Golkar. One of the men who had been a *santri* at Al-Fulan under the founding *kyai*, Kyai Fulan, said that Fulan was dedicated to the Islamic Party (PPP) and would not support Golkar because it would compromise the "holiness/purity" of a *pesantren*. He said that, in 1995, *santri* who wanted to vote for PPP (the only Islamic party during the Suharto era), left the *pondok* (Al-Fulan). They said that in the mid-1990s, in order to grow, many *pesantren* supported Golkar, and hence received many government grants. They repeated several times that, for the most part, the population living around Al-Fulan did not agree with this collaboration.

A public protest over the alleged support of several *kyai* for Golkar further illustrates how *kyai* are evaluated according to their politics. The event also illustrates that such criticism cannot be brought into the public sphere with impunity. On June 11, 1995, a group of student protesters calling themselves the *Santri* and University Student Forum of Jombang (Forum Santri dan Mahasiswa Jombang), most of whom were students at Universitas Darul Ulum, part of the educational efforts of Pesantren Darul Ulum, staged a demonstration protesting a news chapter report claiming that all the *kyai* of Jombang were in agreement that they would support Suharto for reelection. Over 200 students went to Pesantren Darul Ulum to speak with the headmaster and to conduct *tahlilan* (prayers, including *Surah Yasin*) at the grave of the founder of Darul Ulum. When the protesters came to the *pesantren*, they found the main gate locked. Hence the demonstration was conducted outside the *pesantren*. In addition to conducting *tahlilan* for the founder and chanting the history of the Prophet (*selawat* Nabi), they read a statement that accused the *kyai* in question of selling out the Islamic community.

This clear and public criticism of the political alignment of several *kyai* brought swift and harsh consequences. Several key protesters were suspended from Universitas Darul Ulum until they formally apologized and recanted. Several *kyai* told me that the protesters had no morals; it is wrong to publicly criticize a *kyai*. What the event clearly demonstrates is that although *kyai* may be criticized for acting in certain ways, there are limits to how such criticisms may be done. However, one of the cofounders of Al-Hikam in Malang observed that the students were actually protesting the degradation of the concept of *kyai*. He argued that a *kyai* is a leader with multiple functions; one of which is to be like a father with great love for his students. The degradation of *kyainess* happens when *kyai* are no longer like fathers to their students. As an alternative, he offered the example of Hasyim Muzadi, who gives advice and love to his students as well as listens to their complaints and suggestions. He argued that had the *kyai* at Darul Ulum been fatherly toward his students, the protest would have been defused.

In 1995, NU was still following the *Kembali ke Khittah 26* movement, which had removed NU from electoral politics completely and returned it to its original focus as a socioreligious organization. In the 1999 election, NU did not revert to being a political party but did support, even if indirectly, PKB, a political party that was placed overall third in the parliamentary elections. Out of this election, Abdurrhaman Wahid was selected as Indonesia's fourth president.

Even in 1995, those who wished that the NU should not be involved in electoral politics were political actors nonetheless. They took the official position of nonaffiliation with any political party so as to be immune from government regulation. The best example of this is Abdurrahman Wahid, who was first elected to the position of general chairman of NU in 1984 and has been reelected twice. In 1984, he led NU in the *Kembali ke Khittah 26* movement that officially took NU out of any and all electoral activities. NU, its functionaries, and its organizations were not allowed to directly endorse or support any party or candidate. This injunction did not prevent functionaries from doing so as individuals but they must not use their position in NU to do so. Gus Dur argues that those who wish to go into politics should go directly into politics and not use NU. Consistent with this, he stepped down from his position when he became the president.

For Gus Dur, *Kembali ke Khittah* frees NU from government interference. He argues that when *ulama* and government are closely affiliated, the role of *ulama* is severely limited. Further, the religious agenda is determined by the state that has goals, which may be incongruent with Islamic teachings. Therefore he argued, NU should emphasize education, proselytism, and social and economic projects over politics. He and other *kyai* still maintain this general position that most *kyai* should not be too closely associated with politicians.

What direction "politics of *kyai*" now takes as a meaning of evaluating *kyai* is unclear, given that the president is a *kyai*. However, Gus Dur has repeatedly urged *kyai* not to run for public office and has repeated the adage, "A good king seeks out the *ulama*; a bad *ulama* seeks out the king." Others have used this adage to suggest that *kyai* should distance themselves from Gus Dur because he was the head of state and hence the king in the adage. They go on to suggest that *kyai* aligning themselves too closely with Suharto is what kept him in power for so long.

The Identity of Kyai and the Identity of the Pesantren World

Ustadh Abdul Gani at An Nur told me how the *pesantren* world's *kiblat*,[5] in a metaphorical sense, keeps changing. He pointed out that the *kiblat* is not a particular *pesantren*, but a particular *kyai*, based on his personal qualities as discussed above.

At first, it was Kyai Hasyim Asyari and later his son Wahid Hasyim. After the death of Wahid Hasyim, Kyai Kholid (d. 1970) at *Pondok* Sitogiri in Pasururan became the *kiblat* of the *pesantren* world and NU. After Kyai Kholid died, Kyai Asad (d. 1993) of Situbondo became the *kiblat*. According to Abdul Gani, both Kyai Kholid and Kyai Asad could become the *kiblat* because they were of the same generation of Hasyim Asyari.

The identity of the *pesantren* world is largely shaped by the collective and individual identities of *kyai*. To the degree that a *kyai* emerges from the masses, as Edy and Tuti said was true of Kyai Fulan of Al-Fulan, they may be said to be intellectuals in the Gramsciian[6] sense. Kyai Fulan had to support himself with a business, in addition to his religious teaching and hence was seen as belonging to the masses (*masyarakat*). *Kyai* who are born into this elite group may be able to establish firm roots in the *masyarakat* and defend their interests. They too might be considered intellectuals in the Gramsciian sense; members of a class that have risen up to defend that class. The degree to which both the category "*kyai*" and individual *kyai* can be reevaluated, criticized, and debated also indicates the extent to which they might fit the Gramsciian ideal of intellectuals. However, the above material shows that discourse critical of *kyai* is largely covert. The degree to which this discourse is public is limited to general statements of the type: "a *kyai* who does such and such (is more concerned with economic well being than religion, for example) is not a good *kyai*." Many who hear such statements know to whom the speaker is referring, but protests in front of a particular *pesantren*, in which they accuse a particular *kyai* or some wrongdoing as in the protest discussed above, are not allowed. In this way, *kyai* are elite and untouchable; only other *kyai* may criticize them directly.

Leaders and Community Identity

The mostly Indonesian literature on *kyai* rather naively presents *kyai* as leaders of the hearts and minds of Indonesian Muslims in an unproblematic fashion. The material considered here shows how followers closely scrutinize their leaders and even debate the basic nature of leadership. Therefore, it might be possible to understand *kyai* as a member of a class that has risen to defend the interests of that class; they are not an elite class or members of one. Rather they are members of the proletariat, the peasantry, and the rural poor, who by means of their religious education are leaders of the poor and disenfranchised. This is very much how *kyai* want to see themselves, as defenders of the poor, as little more than poor themselves. Taking this view we can understand the material in this chapter as the mechanism by which a class promotes certain persons as members who are just a cut above, who can lead the class/community.

The problem with this view is that *kyai* are highly respected and venerated, in public discourse at least. And for many of them, great spiritual power is attributed to them and from this they garner much adoration. In many ways they are a religious elite with great social and political influence as well as having incomes higher than many of their followers. *Ulama* families tend to marry among themselves thereby creating a largely endogamous group. They are more than a cut above in many respects. Although the norm that *kyai* should be sociable suggests that they should not be socially above common folk.

The idea that *kyai* are an elite class puts this material in a different light. In means that the material reflects, in part, hidden transcripts and counter hegemonic discourses. Is it really an either/or question or can both processes be going on as part of the community identity construction and contestation.

Of the Pope, Victor Turner says "status is acquired through the stripping of worldly authority from the incumbent and the putting on of the meekness, humility, and responsible care for the members of the religion, even for all men" (1969:195). So we have the idea of a leader that comes from the "upper half of the social cone" who then takes on the poverty and humility to become equal to all so that they can be more effective as a leader. So very much like the claim that the leadership comes from below, it is an attempt to cloak authority in the aura of communitas—and given Islam values the equality of believers before God, this makes sense.

It is important to ask whether anyone ever uses this discourse to question the structure itself or like the cargo cult prophets, do they reject certain men's claims but not the basic ideas behind them? It seems that certain men are judged as not really being *kyai* but the basic category is sacrosanct, even for Reformists who generally reject their authority. I have had Reformists tell me with all sincerity that they knew of a *kyai* who flew back and forth to Rome in a matter of minutes to get the original Bible to prove to a Catholic priest that the Bible is mistaken. These were people who rejected the authority of *kyai* but still believed that some could do some pretty amazing feats.

Leadership and Group Identity

I have suggested that the debates about leadership are part of the construction and contestation of a community's identity. Now, I want to make explicit what the material presented here says about the contested identity of the Classicalist Islamic community in Java.

The institutional minimum of having a *pesantren* shows how the community is centered around the institutions of education and proselytization. Further, the curriculum of *pesantren* tells us more about the values of this community. *Pesantren* education balances normative and mystical Islam. It also balances traditional education with modern, "Western" education.

The concern with the mystical powers of *kyai* reinforces the strong Sufi underpinnings in this community. The mystical practices of Sufism was attractive to the Javanese who were already practicing tantric forms and continue to be attractive to many Javanese both in urban and rural areas. One central dividing point between Reformists and Classicalists in Java and elsewhere in Indonesia is place of mystical practice.

Frequently people quoted to me a hadith that states that one should seek knowledge even if one has to go to China (i.e., endure great hardships). Because the Prophet highly valued knowledge, Islam places a high value on those who possess it, especially religious knowledge. The interest of Java in esoteric knowledge reinforces the centrality of the mystical experience for this community.

The issue of morality, however defined, seems to be one of the most important factors for determining the legitimacy of a *kyai*. The moral standards to which *kyai* are held are far higher than those that apply to ordinary members of the community. In the context of corrupt political leadership, this is a strong demarcation between this community and others whom they see as less pious. It was in the area of morality that I was able to collect the most detailed and extensive data. This suggests that the question of morality is central to defining the community identity.

The politics of *kyai* are the most open and confusing part of the material considered here. Neither in 1995 when I first did my research nor in 2000 was there any consensus about what kind of political position a *kyai* should take. It seems simply that a *kyai* you admire and support should share your political views. However, it is clear that some people falsely attribute their own political position to particular *kyai*. What is clear from this material is that the Classicalist community in Java and elsewhere in Indonesia is far from being politically unified. Nowhere else is this more clear than in the community's evaluation of the Abdurrahman Wahid presidency.

Contesting Leadership

In the summer of 2000, when I returned to Indonesia to investigate the interplay between *kyai* leadership and democratic leadership, there was cynicism toward previous uses of traditional models of leadership in the presidency and a strong desire to move toward an open, transparent, and noncorrupt government. In this context, we can explore the interaction of two models of leadership that met in the person of Abdurrahman Wahid: *kyai* and president. A particular interest is how the models are contested; how different groups understand these two concepts, and how they may or may not combine fruitfully. There are dimensions of the *kyai* leadership model that are incompatible with democratizing—the idea that a *kyai* is a "little king" of his *pesantren* and his followers. The process of democratization has also started to change ideas about *kyai* leadership including demands for fiscal accountability. This section will speak to shifting notions of leadership in times of political, economic, and social crisis.

The contested nature of the *kyai* leadership model involves five issues when applied to the Gus Dur presidency. First, is the question of whether or not Gus Dur really is a *kyai*. Second, and alternatively, is he just manipulating the symbolism of *kyainess*? Third, if he is a *kyai*, then he is both a religious and political leader and hence a Caliph. As such, he should meet these requirements. Fourth, is a question rising from the *pesantren* community about whether or not he should have used his *kyainess* while he was president. The wider issue implied in the above question regards the interaction between *kyai* leadership theories and democratic leadership theories.

Is Gus Dur a Kyai*?*

This issue is one that divides along obvious lines. Those who support Gus Dur answer in the affirmative, while his opponents answer in the negative. The division is also along Classicalist/Reformist lines. His supporters (in this case all Classicalists) attribute both his oddities and his greatness to his being a *kyai*. His opponents (again in this case, all Reformists) either question the entire category of *kyai* or they question whether or not he can legitimately be considered one.

Gus Dur is frequently criticized for being inconsistent and confusing. One national NU leader explained these inconsistencies as being due to the fact that he has three intelligences (*cedersan*): intellectual, spiritual, and emotional (*hati-nurani*). Because he shifts between these three very different ways of knowing, he might seem inconsistent to some. The NU leader explained the sudden dismissal of Laksamana Sukardi in May 2000 as Minister of Investment and State Firms by stating,

> Using his spiritual ability, he saw Laksamana and saw that there was something wrong with him and fired him. Some people say that Gus Dur has some ability to predict the future, and talk to spirits. I will neither affirm nor deny this.

Other forms of esoteric knowledge are attributed to him. A number of different stories are told about him locating lost graves of saints. In one story, Gus Dur was approached by a group of people who were concerned about their *kampung* (village, quarter). They sought a solution to the social and moral bankruptcy that they felt plagued their neighborhood. Gus Dur visited the *kampung* and quickly pointed out that the grave of the important and powerful *kyai* who founded the *kampung* was not receiving its necessary care. In fact, no one had even remembered where it was. Gus Dur reportedly told them to take care of the grave and thereby goodness would return to their *kampung*. In another story, he invited a number of *kyai* to go visit a Wali's grave, but the other *kyai* were completely unaware that the grave existed. This tale not only claims that he has this kind of spiritual knowledge, but also that his is superior to that of other *kyai*.

Political opponents who do not recognize him as a *kyai*, do so in part because they might not have belief in those sorts of things at all. Other opponents of Gus Dur wish to undermine his authority as a *kyai* and question his educational preparation. The General Head of Dewan Dakwah stated,

> I have heard that when Gus Dur was studying in Iraq he was not studying religion but socialism and the Bahk party, and bahk means kebangkitan [implying that PKB is a socialist party]. When he was in Egypt, he also was not studying religion but foreign languages. (English, German)

In the context of a *kyai* president, some of President Abdurrahman's opponents wish to contest the exclusive right of Classicalists to use the title *kyai*. The Secretary General of Partai Bulan Bintang argued,

> The title *kyai* comes from the people. In my opinion, *pesantren* people do not have exclusive rights to the title *kyai*. Whoever is knowledgeable

in religion and can live out life in the appropriate way can be called *kyai*. But what the people really need is consistency. If he says he is a *kyai* but his behavior is not what is expected by the people they are going to be disappointed. The people are already disappointed with Gus Dur. They ask, how can a *kyai* want to allow the communists back? How can a *kyai* be inconsistent? How can a *kyai* engage in such shrewd political maneuvers?

This is tapping into the discourse on the morality of *kyai*. By raising such important moral issues, the speaker was questioning Abdurrahman's status as *kyai*.

Just Manipulation?

The question arose whether Gus Dur was using the symbols of being a *kyai* without believing in them. The secretary general of Partai Bulan Bitang (The Star and Crescent Party) argued that *kyainess* (*kekyaian*) is not very important for Gus Dur and that he just uses *kyai* symbolism to gain legitimacy in the eyes of his constituents. He further argued,

> Gus Dur uses his *kyainess* but in truth he engages in rational political maneuvers. He just uses his title *Kyai* to justify his political tactics. I have met with him several times and each time he refers to his prophetic dreams. But in politics he is very rational.

Another opponent of Gus Dur, the secretary general of Dewan Dakwah argued that the claim that Gus Dur is just using *kyainess* for political ends is too vulgar. However, he suggested that many of Gus Dur's followers hold him in a personality cult and even believe that he has supernatural powers. Gus Dur, he averred, exploits their beliefs.

NU *kyai* uniformly rejected these claims. A *kyai* from Pesantren Krapyak stated,

> As to those who say Gus Dur is using his *kyainess*. I say no. His *kyainess* really shapes him (*sangat mejiwai*). Gus Dur's leadership is very open. The problem is that many observers are partisan and involved and not objective.

He went on to assert that those who make such claims are simply "not in touch with reality."

Caliph Standards

Related to *kyai* leadership is the use of leadership standards from the classical texts. In these texts, one finds the requirement that a leader may not be disabled. What is most interesting in this debate is that "Reformists," namely the leadership of Dewan Dakwah who otherwise say that religious authority/knowledge should come from Qur'an and hadith only, call on *kitab kuning* to argue that Gus Dur should not be president,

> I did not nominate Gus Dur for the presidency because he is an invalid; he cannot see perfectly, although a leader is required to have good vision. A leader must be healthy both physically (*lahir*, lit. externally) and mentally, spiritually, and emotionally (*batin*, lit. internally). Although these qualifications are found in *kitab kuning* and not in Qur'an and Hadith, we are not disconnected (lepas) from the majority of *Ulama* who are the inheritors of the Prophet.

On the other hand, NU *kyai* argued that either such requirements did not exist or simply did not apply. Nafik, a *kyai* trained at Gontor argued,

> As far as the *fikih* requirements, we must remember that Gus Dur was elected democratically, which is itself something that is unpredictable. At that time there was still a strong influence felt from Suharto and Habbibe. Certainly from *fikih* discourse he is not allowed, but it happened.

A *salaf kyai* from Pesantren Krapyak argued that there are opinions that if the disability does not impair his duties then it does not matter. He further argued that blindness does not really incapacitate a president. Deafness, on the other hand would mean that a leader could not do his work, which he implied to be listening to diverse arguments and discerning the best course of action. Another *kyai* argues that these notions of leadership that many dimensions of political *fikih* are outmoded including inherited positions of leadership, which clearly do not apply in a democracy. Others argued that as long as the leader's heart was not crippled, other disabilities are insignificant.

Democratic versus *Kyai* Leadership

I now turn to what seems to be the key debate: what kind of leadership model should Indonesia use? Should it use a democratic model? In which case, spiritual explanations and *kyai* style leadership may have little place. Or does Indonesia need spiritual leadership? How did the presence of a *kyai* president shape democratic notions of leadership. Conversely, how did it affect the *kyai* leadership model? Did the pressures put on Gus Dur to act less like a *kyai*, democratize the *kyai* leadership model in general?

Separate Roles?

Gus Isyom, Gus Dur's nephew and a promising young *kyai*, argued that Gus Dur should not position himself as a *kyai* but as a president during the term of his administration. There are aspects of his *kyainess* that cannot be separated from his presidency. For example, Isyom explained,

> When Gus Dur and Megawati first started out in their new offices; they did *ziarah* to Blitar and to Tebu Ireng. *Ziarah* is part of culture of NU and the culture of Java. And this is OK.

But, Isyom argues, Gus Dur cannot not justify his decisions by claiming that while performing *ziarah*, the deceased appeared to him and gave him direction. Rather, he must use rational justifications.

Nafik argued that it is not really possible to really separate Gus Dur as a *kyai* and Gus Dur as a person who has incredible capabilities. He continued,

> He can speak a number of foreign languages. He read *Das Capital*, in German, when he was in junior high. We cannot separate Gus Dur *kyainess* from his very nature and character as a person.

Nafik went on to declare that Indonesia needs him in both of his roles—religious and secular leader. By using both his roles, Gus Dur can be a culture broker,

> We need some one who can lead where we are at. And because the world is the way it is we need someone who is very intellectual. So his *kyainess* is needed because we still think [mystically] but his intellectualism is needed because the rest of the world thinks [rationally].

Kyainess *and Democracy*

Gus Isyom expressed concern about this emphasis on Gus Dur's mystical abilities, namely, that it will interfere with the democratization process,

> There are those who believe he is using dreams and ilmu laduni [esoteric knowledge] and that he has mystical powers. These explanations are common enough in Javanese theories of power (*kekuasaan*). But how are we going to become a modern democracy if these very primitive things continue. To be a modern democracy, is not necessary to leave behind these traditions. But at the very least the president has

to give rational explanations for his decisions. But if for example, I am president, and I fire a minister I cannot give as a reason that I had a dream. I have to have a rational explanation.

Isyom is generally recognized as the only descendent of Hasyim Asyari who is able to return *salaf pesantren* education back to Tebu Ireng, the family *pesantren*. And yet, he is very concerned with democratic ideals.

Nafik's concern was that Gus Dur compromises too much with the presidential system. This is why, according to Nafik, Gus Dur's cabinets cannot work effectively. He stated that this tendency to over-compromise comes from his being a *kyai*, that is, an educator. Nafik averred,

> [Gus Dur] thinks everyone is essentially good and that all can be invited to learn, to compromise, and to be friends. He is an educator, the teacher of the people. (*guru bangsa*)

A key issue to consider is the use of competing models of leadership. Since the founding of the Republic of Indonesia, the presidency has been surrounded by traditional mystical and magical symbolism of one sort or another. Sukarno used Ratu Adil, or Just King, symbolism. Suharto used Javanese kingship theories and symbols. Since the decline of the Suharto regime, many of my consultants have reported a desacralization of the presidency. There is disagreement over whether Habbibe started this process of desacralization or if this honor belongs to Gus Dur. Regardless, some argue that Gus Dur is desacralizing the office, whereas others are concerned that his being a *kyai* is resacralizing the presidency. Gus Isyom expressed the following concerns:

> Suharto used Javanese kingship theories and symbols and it looks like Gus Dur is using *kyai* theories and symbols. When Suharto was president, the presidency was sacred. Habbibie desacralized the presidency. If Gus Dur uses *kyainess* too strong he will resacralize the presidency. On the one hand a *kyai* president may well mean the resacralization of the presidency but on the other, it may lead to the desacralization of the role "*kyai*" Kyai usually only deal with religious issue but at the moment he enters into the center of political power (pusat kekuasaan), he has a responsibility to be transparent. And he cannot claim religious authority.

Another issue raised in the above quote is that Gus Dur's dual role may desacralize or democratize the role of *kyai*. An East Java regional

NU functionary recognized this possibility when he stated,

> But there is starting to be the realization that *kyai* are just people and must be corrected if they are found to be mistaken. This has been done already by other *kyai*. Giving correction to a *kyai* must be done in a very polite and sophisticated manner.

The idea of giving correction to a *kyai* also suggest the insertion of democratic ideals into *kyai* leadership. This seems to be a relatively new development. Whereas as in 1995 it was not acceptable for *kyai* to be publicly criticized as the example of the student protest showed, in 2000, a group calling themselves the "10 special *kyai*" (*sepuluh kyai khas*) formed to be a watchdog group and to send letters of advice and correction to President Abdurrahman on as-needed basis. Granted, in 2000, the idea is still that only *kyai* may criticize other *kyai*, but this in itself is an opening up of this particular institution.

Another example of the emerging the democratization of *kyainess* was seen in a May 2000 meeting of the Jakarta NU chapter. In this meeting, a discussion arose about the need for fiscal accountability structures, particularly around *zakat* contributions. The argument was that these contributions should no longer be handled by individual *kyai* according to their own inclinations without any reporting mechanism or accountability structures. Instead, *zakat* should be given officially to NU chapters, which should use complete and modern bookkeeping methods to generate reports back to the regional NU offices.

Finally, one *kyai* recounted how he warned his fellow *kyai* that when Gus Dur was the head of state that they need to distance themselves from him. He argued,

> Gus Dur is now the president, he is the power (*kekuasaan*), we need to be careful about defending the "powers that be" because we will legitimize them and this is the practice of Suharto era to gain legitimacy from *kyai*.

This idea is supported by the axiom that says a good king seeks out the *ulama*, but a bad *alim* (singular of *ulama*) seeks out the king. Or as Al-Ghazali states

> Another sign of the learned man of the hereafter is that he keeps himself distant from the ruling authorities and avoids their company, because this world is sweet, ever-new and its bridle is in their hands. (1991 I:66)

There were also some antidemocratic trends among *kyai* during Abdurrahman's presidency. Now that NU finally had a significant role in politics, many *kyai* expected to draw on traditional patron–client relationships and benefit financially and politically from having "their man" in office (Barton 2002:303; van Bruinessen 2002).

Discussion

To extend our understanding of the negotiation of modernity and tradition in Indonesian Islam, we have examined the ways in which *kyai* leadership is evaluated. It is hard to determine whether these standards are new. However, my assumption is that leadership is always contested and subject to evaluation and so the process is not new and many of the standards have also long been in place. The standards by which *kyai* are evaluated establish a kind of tradition, namely, the values by which the community wishes to be lead. Other standards, such as political orientation, clearly show the ways in which traditional values interact with the demands of a nation-state.

As much as the election of Abdurrahman Wahid to the presidency in October 1999 surprised some observers, his ouster in July 2001 was easily predicted. It was not my task to unpack the various allegations against Gus Dur or to evaluate the debate about whether his impeachment was constitutional. My goal has been to explore the debates around the Gus Dur presidency from a cultural and religious point of view. As such, some of the key elements discussed in this chapter cast light on his ouster.

There are other ways in which Abdurrahman's presidency was evaluated. Martin van Bruinessen (2002) discusses the attitudes of NU *kyai* in primarily political terms. In sum, they were primarily concerned with the implications for *Kembali ke Khittah*. Many were disturbed by the decision to get out of "practical politics" in the late 1980s and saw Gus Dur's time in office as a return to political participation. The problem that remains is whether NU members owe strict allegiance to PKB (The National Awakening Party), which by all accounts remains Gus Dur's party, or do they remain free to participate with any party they choose. And do they remain restricted from using their position in NU (if they hold one) to promote any political party.

Greg Barton (2002) avers that the criticisms about Abdurrahman's presidency were about practical issues—corruption, Islamic radicalism, an unbridled economic crisis, and poor management—but not spiritual issues. Although there may have been a number of reasons why people opposed Gus Dur, we cannot overlook the fact that many

Reformists opposed him on the grounds of his Sufi orientation (Howell 2001:703), which was expressed in his identification with the *kyai* leadership model, his use of esoteric knowledge, and his visitation to graves. The concerns over his possible involvement in two financial scandals underscore the emerging concern for transparency in government and the desire to abolish corruption. Corruption had been such a part of the Suharto era, that many Indonesians may have hoped that a *kyai* president would have had the financial morality to avoid such scandals. If his ouster was not a victory for democratic models of leadership, it was, at the least, a defeat for the continued use of sacred models in the Indonesian presidency. Third, the relationship between religious orientation and political orientation remains complicated. This chapter has also shown how religious groups who are otherwise different draw on similar symbol sets to draw different conclusions. This shows how the use and understanding of tradition is fluid. People and groups who have, in general, rejected certain traditions, namely the *kitab kuning*, drew on them to reject Wahid's legitimacy. Others who generally draw on these traditions, read around them to support Wahid's legitimacy.

Finally, we must read the debates about *kyai* leadership and part of the general process by which Classicalist Muslims are negotiating modernity and tradition. The debates about the criteria to use to evaluate a *kyai* show that not only is there a tradition about this kind of leadership, but also that this tradition is constructed and contested. Further the debates about Gus Dur's presidency show its relationship to modernity: the need to imagine an Indonesian democracy complete with transparency, accountability, and rule of law.

Chapter 6

A Peaceful Jihad in a Globalizing World

I was chatting with a few students at the Malang branch campus of the State Islamic Institute (IAIN). I was a bit surprised when one of them challenged me, "What do you think of the Huntington Hypothesis?" My surprise did not come from being challenged but from being asked about a hypothesis about which I had not heard. Somehow, in the year that I spent preparing to depart for Indonesia, I had missed Samuel Huntington's *Foreign Affairs* article (1993), which elucidated a theory that since the end of the Cold War we have been moving toward a Clash of Civilizations in which the major poles were the West and Islam. What *is* telling is that they had not missed it; they had been carefully considering these issues and been following the intellectual developments. And not only these students, but also most Indonesian Muslims that I met, were concerned about the relationship between their nation, their faith and the West, modernization, and globalization.

In much of contemporary Western scholarship, modernization is seen as being particularly problematic for Muslims. For example, Daniel Pipes, a leading public intellectual about Islam (see *danielpipes.org*) has written,

> To escape anomy [*sic*], Muslims have but one choice, for moderniza- tion requires Westernization. . . . Islam does not offer an alternative way to modernize . . . Secularism cannot be avoided. Modern science and technology require an absorption of the thought processes which accompany them; so too with political institutions. Because content must be emulated no less than form, the predominance of Western civilization must be acknowledged so as to be able to learn from it. European language and Western educational institutions cannot be avoided, even if the latter do encourage freethinking and easy living.

Only when Muslims explicitly accept the Western model will they be in a position to technicalize and then to develop. (1983:197–198)

Here Pipes, suggests that Muslims must either embrace both modernization and Westernization, or engage in jihad, or a clash of civilizations, against the West. Likewise, Clifford Geertz did not hold much hope that *kyai* would be able to provide both modern secular and traditional religious education and by so doing remain in a position as culture brokers (1960b:249).

In a myriad of ways, *kyai* are proving both Geertz and Pipes wrong and are imagining a modernity that although potentially dangerous can be made Islamic. Through their various activities they are negotiating both tradition and modernity because although modernization does not require Westernization, it does require the (re)invention of modernity and tradition.

Classical Islam Negotiating Modernity

Are *pesantren* people modern? Using the institution-based definitions of modernization theories, they most certainly are. Modernization theorists looked at markers such as technology use, labor specialization, financial institution, political interest groups, and nationalism over group interest. *Pesantren* have incorporated technology in both curriculum and methods. Apart from the simple use of public address systems to preach and make announcements within a *pondok*, telephones, computers, the Internet, automobiles to go to meetings and preaching dates, airplanes to go to Mecca, *pesantren* teach their students how to use these things. Concerning the specialization of labor, secular education provides students the basic skills and socialization needed to be trained in specific labor fields. My discussion of banks in Tebu Ireng indicated some of the critical importance of banking to *pesantren* people. They seek to create a banking system that is not usurious but can meet the financial needs of a developing country.

Pesantren people are involved in a myriad of interest groups that increase their participation in national debates. Chief among these is Nahdlatul Ulama (NU), however, it should be remembered that NU is an umbrella organization and that there are many other interest groups under its aegis. *Pesantren* people also participate in interest groups outside of the NU organization and culture.

The concrete things that *pesantren* people have done to shape Indonesian national character include fighting for Indonesian independence, participating in the early debates about the nature of the

Republic's constitution, insisting that "belief in God" was the first point in the national ideology, insisting that all Indonesians receive a minimal level of religious education, and stressing that the government assist Muslims with performing the *hajj*. In addition, the *pesantren* world has contributed to the establishment, and maintenance, of a secular state. The existence of a secular government has created great difficulties in other Islamic countries (e.g., Iran, Egypt), however, in Indonesia the secular state and the Islamic community have reached an accommodation, as was discussed in chapter 3. The withdrawal of NU from politics was in part in support of the secular state, not of the current regime, perhaps, but support of the idea of a democratic state that represents all Indonesians. A man with Abdurrahman Wahid's pedigree and charisma could easily lead a movement to establish an Islamic state, but instead he was a key leader in the pro-democracy movement and many *kyai* were active in trying to keep his presidency democratic.[1]

Most *kyai* now focus their efforts on creating an Islamic society rather than an Islamic state. This is congruent with the practice of Islam, as a whole, in Indonesia, in which there is greater focus on personal piety than on the external enforcement of religious law. The hope of many *kyai*, then, is to raise the level of personal piety of many Indonesians to such a level as to make a pious nation, and even a pious government, from the bottom up.

One of the chief ways for this community to negotiate modernity has been through education. Abdurrahman Wahid argues that *pesantren* need to be able to make two contributions to society: workers who have *pesantren* morals and ethics, and *ulama* who can engage in a globalizing, technologically oriented society. Many agree that *pesantren* must deal with the changing world and prepare their graduates for college and the workforce. To do any less is a disservice to the students. Further, most parents, especially from urban and urbanizing areas, will not send their children to a *salaf pondok*; they want their children to be able to succeed. Therefore, these *pesantren* have not only government recognized *madrasah* system schools but also have National system schools as well. Some *salaf kyai* look at modernizing *pesantren* and say that if the religious education drops to less than 50 percent, there is little purpose in parents sending their children there or even calling the school a *pesantren*. It should be noted that large *salaf pesantren* still exist and many in the Classicalist Muslim community still value traditional education and want their children to be trained in that fashion, rather than in secular subjects.

In *salaf pesantren*, the whole day is dedicated to religious training. Accepting government education cuts the day in half. Hence, *pesantren* are jealous of this time and try to reclaim as much as possible for religious training. Some argue that the change in focus of *pesantren*, from creating *ulama* to creating intellectuals and workers who are religious, has created a crisis in leadership. Where will future *kyai* come from? Since not all students at *salaf pesantren* have what it takes to become *ulama* some people have argued for a system that tests students after they finish *Tsanawiyah* and places those with the necessary skills in a special *ulama* training program.

This accommodation allows that not all *pesantren* graduates will have the knowledge of *kyai*. The one point that no one in the community is willing to concede is that graduates should have the morality of *kyai*. Further, teaching the basics of Classical Islam based in the *kitab kuning* is seen a way to prevent the growth of Islamist extremism.

Many *pesantren* people associate the processes of modernization and globalization with the loss of traditional values. Nafik of Al-Hikam said that this happens mostly because many naive people link Westernization and modernization, a linkage Howard Federspiel (1996:202) attributes to the writings of Siradjuddin Abbas. Nafik argued that much of what is done in Indonesia today is Westernization without any real modernization. Education can overcome such naiveté and hence, he says, the goal of Al-Hikam is to train modern people (arts and sciences students) with traditional values. People so trained will be able to lead the nation so that it can engage in globalization and forge a new national identity consistent with an (imagined) Islamic heritage.

Several people mentioned that giving up Islam is not necessary for modernization. This very claim assumes a modernity in which the spiritual is challenged. Indeed, Abdul Gani, an *ustadh* at *pesantren* An Nur argued that "man-made religions" like Buddhism and Hinduism were incompatible with modernity. On the other hand, he argued, Islam, as a revealed religion, is good for all times and can fit with modernity. However, elements of popular Islamic practice need to be excised from the communal body of practice in order for Islamic countries to prosper. He identified these as aspects of popular mysticism (*kepercayan*) around ancestral spirits. He clearly supported the notion of the place of Sufism in modernity, as did many others.

Defending the Peaceful Jihad: Negotiating with Radicals

In recent years, the processes of Islamization, globalization, and democratization have allowed expressions of Islam not previously witnessed in the country. For example, the emergence of radical groups like Laskar Jihad and Jemaah Islamiyah (JI) in the early to mid 1990s. Large-scale acts of terrors and the "War on Terror" have forced moderate Indonesian Muslims to negotiate their identity vis-à-vis small minorities of Islamic radicals.

Laskar Jihad was a group based in Central Java that recruited and deployed militia to an area in Eastern Indonesian where there was Muslim–Christian conflict. The net result was an exacerbation of the violence. Its founder Jaffar Umar Thalib is a veteran of the Afghanistan jihad against the Soviets. Beyond that there seems to be no connection with global terror networks. The organization officially disbanded in 2002.

JI is linked to Al-Qaeda but is not controlled by it. The relationship between the two organizations is multifold: overlapping membership, shared operations planning and execution, including the September 11 attacks. JI was founded by two Indonesian radical clerics, Abu Bakar Ba'syir and Abdullah Sungkar, both of whom were in Malaysia living in exile. Other major figures associated with the creation of JI are Riduan Isamadduin, also known as Hambali, and Mohammed Iqbal Rahman, also know as Abu Jibril. Hambali and Abu Jibril were alumni of the Afghanistan jihad and had been recruited into Al-Qaeda. In 1994, they were tasked with creating a network in Southeast Asia. Although founded in the early 1990s, JI's first terrorist acts were committed in 2000.

I returned to Indonesia in May 2000 with an interest in how the mainstream Indonesian Muslims, including the *pesantren* community, were responding to Laskar Jihad and other radical groups. An interesting question is that if the majority of Indonesian Muslims are not radicals, why was Laskar Jihad allowed to walk the streets brandishing, at least symbolic weapons, collect money on the streets of Java, and distribute literature. I interviewed a number of leaders from NU and from other Muslim organizations and also listened to a number of debates and speeches during this time. In the context of increased Islamic radicalism, there was great interest in carefully defining one's own identity vis-à-vis radicalism. Some Muslims were sympathetic but not supportive; others were opposed to Laskar Jihad's basic charter. This material allows us to consider the *pesantren* world vis-à-vis other Muslims in Indonesia.

Afandi Rihwan, the interim general head of Dewan Dakwah, a fundamentalist organization, stated that he did not agree with Laskar Jihad but argued that Muslims had the right to defend themselves and to render assistance to other Muslims under attack. He asserted that the assistance being rendered by Dewan Dakwah was strictly are sending humanitarian aid namely rice, clothes, and medicine. Rihwan made it clear that although he did not agree with Laskar Jihad's use of violence, he understands it. However, his colleague, Hussein Umar, the secretary general of Dewan Dakwah denied that Laskar Jihad was engaging in violence. He argued that jihad can have three meanings: (1) protests in the street; (2) trying to correct the government; and (3) war. He further averred that Laskar Jihad had yet to engage in the third type. He argued that their actions were primarily in trying to put pressure on the government because of their deep disappointment with the government's actions in the region. And when they asked permission from the government to go to Ambon, it was to give humanitarian aid.

His insistence that Laskar Jihad had not broken any laws or encroached on anyone's human rights was telling. Shortly after I interviewed him, I collected several copies of a flier called *Maluku Hari Ini* in which Laskar Jihad recounted how many Christians had died by their actions. The denial by Dewan Dakwah of the violent actions of Laskar Jihad can be seen as an attempt to shape Islamic identity, especially in regards to how it is viewed by outsiders. It is unclear whether the Dewan Dakwah leadership was indeed unaware of the violent action of Laskar Jihad or whether they were trying to keep me from being aware.

Typical of responses from the *pesantren* community, Nafik, the day-to-day director of Hasyim Muzadi's *pesantren* Al-Hikam questioned whether the term "jihad" really applied in this case. He argued that there seem to be factors other than religious ones at stake. He argued that, given a centuries-old tradition in Ambon of Christians helping Muslims build mosques and Muslims helping Christians build churches, we must ask whether "invisible hands" had been involved. He, and others, suggested that Laskar Jihad had the backing of certain military factions and even received field training from them. This theory holds that the military might be using such violence to make a case to return to its former involvement in domestic affairs.

Since jihad is an important part of Islamic thinking, defining how to conduct jihad is an important part of defining the Islamic community. I asked a group of *santri* at Al-Hikam, what they thought about Laksar Jihad and they said that Laskar Jihad were far too extreme and reminded me that there are many ways to conduct jihad: social efforts, education, and humanitarian aid. They concluded that violence just is not required. Further they rejected the claim that Laskar Jihad was only involved in humanitarian aid and recognized that organization for what it was. They argued that only a few Muslims feel that there is a need to defend Islam using violence and that such violence can destroy Islam.

In June 2000, Hasyim Muzadi, the General Chairman of NU was ready, pending governmental permission, to send NU's security group (*Banser*)to Maluku to fight a jihad in defense of the Christians. This is a clear rejection of Ustadh Jaffar's interpretation of Laskar Jihad actions as jihad. Further, Dr. K.H. Sayyid Akisyarat—acting as a representative of the NU central offices—addressed a *haul* attended by over 500 people in June 2000 favored a peaceful understanding of jihad, he stated

> Jihad is not violence. Except in a few cases, violence is not justifiable. Jihad can also mean struggling to make sure the Indonesian government meets the needs of its people, both Muslim and non-Muslim. This means subsidies for rice, oil, clothes; so that people can afford these necessities. This kind of jihad will create a prosperous nation. Burning churches will only ruin our nation. Burning churches is not right according to NU. We have to end violence between religious and ethnic groups.

Although Dewan Dakwah's discussions of the definitions of jihad seem to have been motivated by wanting to obscure what Laskar Jihad is doing, NU people clearly described what Laskar Jihad was doing and denounced it.

Laskar Jihad's parent organization was Forum Komunikasi Ahlus Sunna Wal Jamaah (FKAWJ). By claiming to be *ahlus sunna*, Laskar Jihad was trying to establish its legitimacy. Since NU had been the foremost proponent of *ahlus sunna*, some wondered what the relationship was between NU and this new group. NU *kyai* spoke out and clearly distanced NU from this organization.

Kyai Taufik Ali of Pesantren Krapyak, Yogakarta made it very clear that NU distanced itself from Laskar Jihad. He stated,

> When the other group started as Forum Komunisi Ahlus Sunna Wal Jamaah, *kyai* got together and declared that it had nothing to do with NU and had no connections, structurally or otherwise.

Another *kyai* suggested that there was little they could do—that their opposition was "cultural." He stated,

> What is clear is that they are not using the concept of ahlus sunna as it used by NU. NU's concept of ahlus sunna is based on: *tawasud, tawwazul, ahl ahdul. Tawasud* means taking being middle road. *Tawwzul*—which is balanced. *Ahl ahdul*—just, fair (*adil*), tolerant (*tassamu*). He argued that it was clear that for right now the Forum is extreme, intolerant, and violent. Hence, NU sought to oppose them creating a cultural

discourse. Anything more, anything physical, he argued, would be illegitimate and would violate their freedoms.

Another way in which some Muslims wished to distance themselves from Laskar Jihad and other radical groups was by seeking to actively work against fundamentalism or radicalism. Hasyim Muzadi said that he founded Al-Hikam to meet the needs of nonreligion majors whose knowledge and practice of Islam was lacking. The goal was to target students with strong religious sentiments but who have yet to receive solid religious training. This group of students is seen as particularly susceptible to fundamentalism and Hasyim Muzadi expressed his desire to prevent the growth of fundamentalism among college students. Fundamentalism imagines the relationship between tradition and modernity as antagonistic (Lawrence 1989).

At Al-Hikam and some other *pesantren*, the teachers and staff are proactively moving to ensure that classical Islam rather than fundamentalism is the basis of the Indonesian Islamic negotiation of modernity. Muslim scripturalists, sometimes referred to as fundamentalists, reject historical commentary on religious issues, that is, the classical texts, in favor of the personal reading of the Qur'an and hadith. Because *pesantren* people maintain the use of the classical texts as a source of religious authority they are not scripturalists. Further, Hasyim Muzadi has stated that the historical interpretations of great scholars are a tempering factor that can prevent extremism.

Will the Real Pesantren Please Stand Up?

Another way, the NU *pesantren* community attempts to distance itself from other interpretations of Islam is by carefully applying the label "*pesantren*" and discussing what the features are of a "real *pesantren*." A real *pesantren* must maintain certain key features: (1) the use of classical texts, at least in translation or in derivative forms, but preferably in the original Javanese and Arabic; (2) character development and moral training; and (3) an emphasis on the balance between normative and mystical Islam.

The less an institution emphasizes religious education and character development, the less likely it is to be considered a true *pesantren*. There are some schools that claim to be *pesantren*, but the NU *pesantren* people that I spoke with refused to acknowledge them as such. This allows them to deny any similarities between them and groups like Laskar Jihad which are ran out of a "*pesantren*."

This points to a fundamental difficulty with how we approach Indonesian Islam. Even though the Geertzian trichotomy (Geertz

1960a) has long been challenged (Hefner 1985; Koentjaraningrat 1963; Woodward 1989), there is still a tendency to engage in explanatory reductionism, that is to use group membership or other common features to explain behavior. In fact, in this book, I sometimes lapse into making broad statements about NU. Common enough is the assumption that all Islamic boarding schools, *pesantren*, share in a *pesantren* culture and participate in the *pesantren* world. And even in the NU-oriented *pesantren* community there is debate about what counts as a true *pesantren*. It does not help to exclude as non-*pesantren* the schools run by Laskar Jihad and Jemaah Islamiyya and still maintain that *pesantren* do not engage in radical Islam. This is an act of desperately trying to hang on to the idea that organizational features or group membership will be predictive of ideology or even of behavior. I propose that new approaches need to be engaged. In particular we need to remember that there is no formal institutional hierarchy and that rather than thinking in terms of unified communities we are dealing with networks. Future research should look at the variations in Indonesian Islamic communities and sketch the contours of the various networks found within it and beyond the borders of Indonesia as well.

Related to the issue of networks, is the question of group identity negotiation. In discussing groups like Laskar Jihad and other radical groups, NU-oriented *kyai* are hesitant to call them non-Muslims but claim that they have little religious education. This becomes a way of defining both themselves and others. However, it is clear that various groups overlap in their use of certain texts and ideological arguments, even when reaching opposing conclusions.

Islam, Modernism, and Globalization

The issues discussed above are not unique to Java and Indonesia; they have been faced by Muslims everywhere. Modernization raises a crucial question for Muslims: can they adopt the technology of the West and still hold fast to the teachings of the Prophet? Or are the values of the West (and Westernization itself) inseparable from Western technology (and modernization), as suggested by Pipes in the quotation at the beginning of this chapter?

Bernard Lewis reminds us that Islamic concerns with modernization and Westernization are not new; they began in the late sixteenth century. Expanding European colonialists had faster ships and more powerful weapons than the Islamic world;[2] this meant that, on the battlefield and in the market, European warriors and traders enjoyed an advantage over their Muslim counterparts (Lewis 1997:117).

Beginning with these encounters, Muslim rulers sought to obtain the technology of the West. At first, this was done by simply buying ships and weapons. However, it quickly became clear that this was an unreliable source and that Muslims needed to manufacture these ships and guns themselves. Hence, at an early date, Ottoman and other Middle Eastern rulers adopted a conscious policy of reform, although they did not call it Westernization. In fact, the literature of this time used several terms to denigrate the very notion: "aping the ways of the Franks" in Arabic (*Tafarnuj*) and "Westoxication" in Persian (*Gharbzadagi*) (Lewis 1997:121). Lewis also reminds us that in an earlier period of "modernization" (in the late Middle Ages) Europeans may have well asked "can we adopt the technology of the Muslims and still hold fast to the teachings of Christianity" as they adopted the Muslim innovations of experimental science, algebra, and astronomy as well as paper, the zero, and positional numbering, which Muslims brought from China and India respectively (1997:129).

Contemporary modernization is historically linked with the West and it is difficult to separate the two in contemporary Islamic discourse. Bernard Lewis argues that since the sixteenth century, there have been three basic attitudes toward modernization and Westernization that Muslims might take (1997). I would expand Lewis' treatment of the subject by suggesting that each attitude is based on a particular imagined modernity. The first attitude identified by Lewis is that of a supermarket: Muslims may adopt what they find useful without adopting the religion or the values of the West. An extreme form of this attitude is found among the so-called Islamic fundamentalists, who imagine modernity as immoral and corrupting. Lewis associates this position specifically with the Ayatollah Khomeni who decried the United States as the Great Satan, or the seducer of Islam. The second attitude imagines a modernity that can have both positive and negative aspects and seeks to marry the best elements of both modernity and Islam. However, Lewis argues, more often than not the result is not a marriage of the best but "a promiscuous cohabitation of the worst" (Lewis 1997:127). The third attitude was that of Kemal Atatürk and the Young Turk movement and imagines a modernity that is not only mostly positive but also one because of its ascendancy requires Muslims to either join with it or be destroyed (Lewis 1997:127).

The *pesantren* community is creating a fourth option, one that imagines an Islamic modernity. At first glance, the *pesantren* world seems to be taking the third attitude. Repeatedly, I heard the argument that *pesantren* must change to deal with changing times; globalization requires new educational strategies. Some *kyai* use a market analogy and suggest that *pesantren* must become competitive or be driven out of the market. In other ways, however, the *pesantren* world seems to be taking the first attitude and would strongly oppose an Indonesian Atatürk; they wish to be selective about what is borrowed from the West. However, *pesantren* people would be as equally opposed to an Indonesian Khomeni because of their commitment to classical Islam. And this is the key difference; fundamentalists are not only selective about modernism, they are also selective about religion. Most Islamic fundamentalists reject the traditions that include such figures as Syehk Abdul-Qadir Gilani and Imam Al-Ghazali, traditions to which *pesantren* people hold firmly.

Pesantren people are doing more than simply trying to marry the best of both worlds, they are making an Islamic modernity. If modernity entails a set of attitudes about authority, time, society, politics, economics, and religion, then the leaders of the *pesantren* world are trying to shape those attitudes. The ultimate concern is still with salvation and the hereafter. Concerns about this world are fine as long as the hereafter is not forgotten. They are aware of the Enlightenment thesis that this world is all there is, and they consciously reject it.

Conclusion

The esteemed sociologist Robert Bellah pointed out that modernity should be seen not "as a form of political or economic system, but as a spiritual phenomena or a kind of mentality" (1968). By defining modernity as a kind of mentality, *pesantren* people have moved the discussion from changes in institutions, about which they may have little control, to matters of the heart and mind. Capitalism, urbanization, and secularization are here to stay. Democratization may or may not continue to influence Indonesia. Keeping a focus on the institutions of the modern world, radical Islamists have only one choice—to oppose modernity. By shifting the focus of modernity to mentality, the *pesantren* community can imagine a modernity in their own image. Of course, they want the technology and the benefits of some of the institutional changes associated with modernity. However, in terms of the mentality of modernism, they wish to define an Islamic modernity. There are certain values and morals they wish to have underpin modernity. These values include Islamic brotherhood (*Ahwuya Islamiya*), selflessness (*keikhlasan*), simplicity in living (*kesederhanaan*), and self-sufficiency (*kemandirian*). Also included is a concern for social justice and serving the needs of the poor.

Taken together these values define a modernity quite different from those practiced in the West. Perhaps the greatest concern *pesantren* people have about modernization is the threat of egoism, or the emphasis on individual gain over communal gain. The values of Islamic brotherhood and selflessness, then, are seen as safeguards to heartless entrepreneuralism. "Simplicity in living" is a control for rampant consumerism, and with the emergence of credit cards, a way to avoid the financial morass in which many Americans find themselves. "Self-sufficiency" gives both the individual and the nation continued independence. For individuals, it means that one should seek self-employment, the very entrepreneurialism that development requires, however, one controlled by Islamic values. For the nation, it means avoiding the kind of metropole–satellite relationship that André Gunder-Frank says creates underdevelopment (1969).

It is in the context of modernization, fundamentalism, and education that we must understand Syehk Abdurrahman's visit to Al-Hikam as related in chapter 1. Al-Hikam, like each of the *pesantren* in this study (each in its own way), is saying that modernization and traditional religious values must be linked. Hasyim Muzadi said that he arranged Abdurrahman's visit so that the *santri* could be trained in *Sufism*. Indeed, this is an important question in Islam—are mysticism and modernism incompatible? Or even, are religion and modernism incompatible? And by providing college students with the opportunity to be both world savvy intellectuals and mystics, the staff at Al-Hikam not only answers this question with a resounding, "no," but they are also taking practical steps to make an Islamic modernism.

Abdul Gani, at An Nur said that modernization, for the West, has meant abandoning religion but that the Islamic world must develop without abandoning Islam. However, he argued that according to Dr. Fazlur Rahman[3] and other Reformists, certain aspects of Islam, especially mysticism, must be left behind. Some of Rahman's writings seem to confirm Abdul Gani's opinion of them. Rahman says that in order to reconstruct Muslim society, Muslims "must take into account the colossal moral and spiritual debris which is the legacy of *Sufism*" (Rahman 1979:244). He further argues that "mass phenomenon of *Sufism*" has "various, deep, and paralyzing effects on society" (Rahman 1979:245).

Many, if not all, in the *pesantren* world strenuously object to Rahman's formulation. Abdul Gani clarified that practices supported by the texts should be maintained; however, he felt that many Reformists, including Muhammadiyah, wish to discontinue practices that have firm support in the Qur'an and hadith. He argues that Indonesia needs both Muslims who can master technology and those who control mystical power (*kekuatan*) and awareness (*keyakinan*).

The *pesantren* case is by no means unique, and although it is beyond the scope of the present study to engage in a detailed comparison of education in other Islamic countries, it is useful to highlight a few comparative cases. In rural Kenya, *pesantren*-like institutions (*madrasa*) emerged in the late nineteenth and early twentieth century as the area was Islamized (in urban Kenya, Islamic educational traditions are much older). Like the Javanese *pesantren*, the Kenyan *madrasa* has adopted general education as alternatives to both government and Christian schools. Also like the *pesantren*, the Kenyan *madrasa* is a center for the defense of popular Islam and a network for community solidarity (Sperling 1993). In Turkey, there has been a strong division between religious education and secular education; the secular government does license some *madrasa* to produce needed judges, but it tries to suppress unlicensed *madrasa*. Otherwise there is no contact, and the *madrasa* do not, and cannot, offer secular education despite the fact that Turkey's public education system is severely overtaxed. According to UNICEF's estimates, 29 percent of Turkish girls are uneducated, compared to 7 percent of Iranian girls (Mater 1996, 1997). Iran's *madrasa* system, as described by Mottahedeh (1985), bears a strong resemblance to the Javanese *pesantren*, or at least to the older *salaf* system. Iran, however like Turkey, kept secular education and *madrasa* education separate. It is from this exclusively religious system that the Ayatollah Khomeni emerged. It would be foolhardy to draw grand conclusions from these vignettes, but they seem to suggest that the hybrid system that has emerged in the form of the *pesantren* studied here will provide Indonesia with far greater educational and religious stability than either Turkey or Iran have experienced.

Huntington and others who predict, or at least fear, an inevitable clash between Islam and the West, in part base their predictions on the assumption that Islam and modernity are fundamentally incompatible. The efforts of the *pesantren* world refute to this assumption. *Pesantren* people are leery of what they see as the deleterious side effects of modernism—materialism, egotism, sexual promiscuity, and indirectly, religious fundamentalism—and they wish to circumvent these problems. They seek to construct an identity that can engage in selective borrowing from the West. This book has shown how *pesantren* shape their curriculum so that their students may gain the knowledge and skills they need to engage modernity. And yet technology is not the end goal of *pesantren*; they are interested in perpetuating classical Islam, which is opposed to fundamentalism. Because *pesantren* people engage in a peaceful jihad and oppose fundamentalism, they pose little threat to the West. This is important because *pesantren* people lead the majority of Muslims in the largest Islamic country in the world. Hence, the single largest population of Muslims in the world is committed to changing society and the world through education, example, and preaching rather than through violent conflict.

Finally, *pesantren* people are carving out a new kind of identity. They reject both an Atatürkian blind embracing, and a Khomenian blind rejection, of all that is Western and modern; they are cautious of globalization and its McWorldian tendencies but nonetheless actively engage it, through the peaceful jihad of *pesantren* education.

However, the most significant question may be whether or not we in the West are willing to think of modernity as a mental state rather than a set of institutions. And whether, as mental states, there might not be mutually compatible, but different versions of modernity. Can we give up our precious institutions that we associate with being modern and democratic? By this I mean, can we accept that there may be other ways to achieve the same goals? Can we accept Islamic versions of democracy, human rights, and civil society as equally good as ours even though they would not work in our societies? Or must we remake Islamic societies in our own image? What I have learned from *pesantren* people is that for a peaceful world, our struggles to improve the world must start with ourselves and then through education and persuasion move out to others.

Notes

Chapter 1 Negotiating Tradition, Modernity, and Identity

1. In 1999, Hasyim Muzadi replaced Abdurrahman Wahid as General Chairman of NU. In May 2004, he was named as Megawati Sukarnoputri's vice-presidential running mate.
2. The Arabic phrase, transliterated in Indonesian, is "Al Muhafazhah Bil Qadimis Shalim, Wal Akhadzu Bil Jadidil Ashalh."
3. Other than a few well-known self-proclaimed Shiites, all Muslims in Indonesia are Sunni.
4. Ibn Tamiya advocated several positions later taken up by "modernists." First is his use of *kiyas* (reasoning by analogy) in reading and understanding Qur'an and hadith. Second, he bitterly attacked the practice of *ziarah*, or pilgrimage to the tombs of saints and called these practices *bida* (innovation). It should be noted that similarities between Ibn Tamiya and Muhammadiyah in Indonesia are not accidental as Ibn Tamiya's teachings were used by the founders of the Wahhabis (Cheneb 1974:152) and Muhammadiyah.
5. The Indonesian government requires all its citizens to claim one of five official religions: Islam, Catholicism, Protestantism, Hinduism, or Buddhism. Many highland peoples have chosen one of the forms of Christianity for their official religion while still maintaining traditional practices. Others have tried to have their traditional religion recognized as a form of Hinduism. This discourse is simultaneous with that examined here and the two influence each other to varying degrees.
6. He includes secular ideologies under the term *faith* because they generally involve value orientations (1991:75).

Chapter 2 The "House" that Change Built

1. Diponegoro was a Javanese prince, affiliated with *pesantren*, who led the Java War (1825–1830) against the Dutch.
2. Indeed, Megawati Sukarnoputri and Abdurrahman Wahid have been major players in the post-Suharto era. Abdurrahman was served as the fourth president from October 1999 to July 2001 when he was impeached and replaced by his vice-president, Megawati.

Chapter 3   "Politics" by Other Means: Using Education to Negotiate Change

1. An interesting study comparing two Muslim cultures is Geertz's *Islam Observed*. Geertz compares how common Islamic concepts, like that of saints, are expressed in very different ways in two different countries.

2. The terms *pondok pesantren, pesantren,* and *pondok* are used interchangeably in the *pesantren* world. In general Indonesian usage, only *pondok* (hut) is used to refer to something other that a *pondok pesantren*, such as *pondok wisata*, or visitor's hut. In contemporary usage, the "*pondok*" in *pondok pesantren*, indicates the presence of a simple dormitory.

3. It should be noted that the term *madrasah* is used quite differently in Indonesia today. At minimum it means a religious school that sits students in neatly rowed desks and uses formal means of examination and evaluation. More commonly it means a religious school following the national curriculum. Therefore I will use the Arabic transliteration *madrasa* when referring to the pan-Islamic cultural category to which *pesantren* belong, and the Indonesian derivative "*madrasah*" when referring to the Indonesian religious day school. It is unfortunate to pin such a large difference of meaning on a single letter, but by following Indonesian spelling conventions, I have no choice.

4. Similar claims are made by other groups. The Jogyakarta royal court, for example, makes this claim because there was a time, just before international opinion went against the Dutch, when the territory controlled by Indonesian forces was coterminous with the Jogyakarta *kraton* (palace).

5. In fact, a group of students assembled for *wetonan* is sometimes called a *halqa* (Arifin 1993:38).

6. Contemporary "wandering students" are quite unlike Kyai Wahab as the archetype of the late nineteenth and early twentieth-century "wandering student." First, while Kyai Wahab obtained the breadth and depth of his learning by studying under multiple teachers, most contemporary "wandering students" are simply supplementing their education elsewhere with a few weeks of study. Second, while Kyai Wahab's wandering sometimes kept him under one teacher for a period of a few years, contemporary "wandering students" spend at most a few weeks under a different teacher before returning to their main teacher.

7. It should be noted that this is a specific appropriation of a general term that means religious school. The usage in the Indonesian context to mean government recognized religious school is hence unique. On the other hand, many *pesantren* have a form of education called a *Madrasah Dinniyah*, which are exclusively religious in their curriculum and are not included in the category of *sekolah*. This form is often found in more traditional (*salaf*) *pesantren* because the amount of time students are now able to spend in the *pesantren* is much smaller than in the past; in part due to the fact that they may seek a general education before entering the *pesantren*. When *Madrasah Dinniyah* are

found in modernizing *pesantren*, it is often as a parallel to the school system, so that the students in this program do not feel different from their counterparts in the *sekolah*. *Madrasah Dinniyah*, sometimes also called *Madrasah Salafiyah*, are not regulated by any government agency. However, NU sponsors *haliqoh* (I.; seminars from A.; *halqa*, study circles) with *kyai* to create a suggested uniform curriculum for all such programs within NU-affiliated *pesantren*. During 1995 this process was just starting.

8. This is a contested term. In Indonesia, *salaf* refers to the practices and ideas set through centuries of legal scholarship. In other parts of the Islamic world, *salaf* refers to the practices and ideas set at the time of the Prophet and the Companions.

Chapter 4   Beyond Education

1. Hasan argues that this hadith should not be interpreted as saying that most Muslims will end up in Hell, because as he argues, most of them are not involved in intentional, divisive innovation. Hasan also suggests that the mention of the Fire does not imply a eternal situation (1994).

2. There are two major branches of Islam: Sunni and Shi'i. The most fundamental division between these two branches is over religious authority and the question of who are the successors of the Prophet. Sunni Muslims argue that the successors should be chosen by election and that religious authority is based on the consensus (*ijma*) of religious scholars (*ulama*). Shi'i Muslims feel that Muhammad chose his cousin and son-in-law Ali as his spiritual and secular heir, and that succession should be through his bloodline; they are hence called "the partisans of Ali." This successor is the *imam*, who is the foremost authority of the law of Islam and a guide for understanding esoteric knowledge (Lapidus 1988:117). Goldziher argues that in ritual and legal aspects, "Shi'i religious doctrine differ from the Sunni in petty formalities that rarely touch essentials" (1981:205). In fact, there are greater similarities between Shi'i ritual and the ritual of the Shafi'i *madhab* than between the Shafi'i *madhab* and the other three Sunni *madhab* (Goldziher 1981:205).

3. In part this was a result of the "long march" of the Siliwangi Division after the Renville agreement. This left Hizbullah groups as the main "national" forces in West Java, although there were some regular TNI units left behind as "rebels" (Arto 1994:114, 122).

4. The Panca Sila are the five principles of the national ideology. They are:

   (1) Monotheism (*Ketuhunan Maha Esa*).
   (2) Humanitarianism (*Kemanuisaan yang adil dan berahad*).
   (3) Indonesian national unity (*Persatuan Indonesia*).
   (4) Democracy (*Kerakyatan yang dipimpin oleh hikmat kebijaksanaan dalam bermusyawaratan, perwakilan*).
   (5) Social justice for all (*Keadilan sosial bagi seluruh rakyat Indonesia*).

Chapter 5 You Are Who You Follow: Negotiating Leadership and Community Identity

1. The term "*nyai*" is also used in Indonesian literature to refer to an Indonesian mistress of a Dutch colonialist. How and why the term has come to have these two widely divergent meanings is unclear.

2. In Islamic creation mythology, *jin* were created before humans but after angels. While humans were created from dust and angels were created from light, *jin* were created from fire (Qur'an *Surah Al-Hijr*:25–27, Lane 1860:222). Some *jin* are Muslims, others are infidels. According to Lane, the non-Islamic *jin* are also called *sheytans*, or devils, the leader of which is Iblis, who refused to pay homage to Adam when he was first created (Qur'an *Surah Al-Hijr*:31–33; Lane 1860:222; el-Zein 1974:175–176). Although *jin* are mentioned in the *Qur'an*, much of what is believed about them is found in a great supply of apocryphal stories about them. Either Muslim or non-Muslim *jin* may bother (*ganggu*) Muslims. Generally the Muslim *jin* will bother another Muslim only if they have been hurt in some way (being sat upon, having something dropped on them, being run over by car). If *selawat* is read often enough, *jin* will be in awe of the reader and not bother them. Reading the *selawat* frequently will also protect someone from black magic (*santet*).

3. It can be noted that urinating while standing is seen an animalistic behavior and belongs to that category of behaviors which are not sins but one receives merit for not committing.

4. This is a pseudonymous *pesantren* and all persons associated with it are also pseudonymous. In this case I have elected to mask the identity of the *pesantren* because of the negative things said about it and its various *kyai*. It is important to document that people sometimes talk about *pesantren* and *kyai* is less than flattering terms. However, knowing which particular *pesantren* was being criticized is not only unimportant but also potentially damaging.

5. *Kiblat* literally means the direction of prayer, of which there is one: toward the *Kabah* in Mecca. Used metaphorically, the term refers to the center of spiritual leadership.

6. For a solid introduction to the philosophy of Antonia Gramsci, see his *Selections from the Prison Notebooks* (1971).

Chapter 6 A Peaceful Jihad in a Globalizing World

1. I cannot explore the complexity of the Gus Dur presidency and the seeming decay of democratic ideals when it was threatened. For a further treatment of the man and his time in office, see Barton 2002.

2. Lewis explains that European countries (especially Northern European) were faced with rougher seas and more bellicose neighbors than their

Islamic counterparts. Hence, it was environmental pressures rather than intellectual superiority that gave the Europeans their advantage (1997:119).

3. Fazlur Rahman was a leading Pakistani reformist who wrote in English and taught at the University of Chicago. Among his pupils was the Indonesian neo-modernist Nurcholish Madjid.

References Cited

Abbas, Siradjuddin. 1969. *I'itiqad Ahlussunnah Wal-Jama'ah*. Jakarta: Pustaka Tarbiyah.

———. 1995. *Empat Puluh Masaslah Agama (Forty Religious Problems)*, 25th printing. Jakarta: Penerbit Pustaka Tarbiyah.

Abdul Quasem, Muhammad. 1975. *The Ethics of Al-Ghazali: A Composite Ethics in Islam*. Selangor, Malaysia: Muhammad Abdul Quasem.

Abdullah, Taufik. 1987. "The Pesantren in Historical Perspective." In Taufik Abdullah and Sharon Siddique, eds. *Islam and Society in Southeast Asia*. Singapore: Institute of Southeast Asian Studies.

———. 1996. "The Formation of a New Paradigm? A Sketch on Contemporary Islamic Discourse." In Mark Woodward, ed. *Toward a New Paradigm: Recent Developments in Indonesian Islamic Thought*. Tempe, AZ: Arizona State University Program for Southeast Asian Studies.

Abu Amar, Imron. 1989. *Sebuah Jawaban Bahwa: Kitab Manakib (Syeh Abdul Qodir Jaelani) Tidak Merusak Aqidah Islamiyah. (A Response that the Manakib of Abdul Qodir Jaelani does not damage Islamic Doctrine.)* Kudus: Menara.

Adas, Michael. 1979. *Prophets of Rebellion: Millenarian Protest Movements against the European Colonial Order*. Chapel Hill: University of North Carolina Press.

Adnan, Saifuddin. 1981. *Arabic Teaching Result in Class V Period 1980/1981 between Ordinary and Experimental Classes at Pondok Modern*. BA Level (Sarjana Muda) thesis. Ponorogo: Insitut Pendidikan Darussalam.

Ali, AKH Ayyub. 1961. "Maturidism." In M.M. Sharif, ed. *A History of Muslim Philosophy*. Delhi: Low Price Publications.

Anderson, Benedict. 1972. *Java in a Time of Revolution*. Ithaca, NY: Cornell University Press.

———. 1990. *Language and Power: Exploring Political Cultures in Indonesia*. Ithaca, NY: Cornell University Press.

———. 1991. *Imagined Communities: Reflections on the Origin and Spread of Nationalism—Revised Edition*. New York: Verso.

Appadurai, Arjun. 1996. *Modernity at Large: Cultural Dimensions of Globalization*. Minneapolis: University of Minnesota Press.

Arifin, Imron. 1993. *Kepemimpinan Kyai: Kasus Pondok Pesantren Tebu Ireng (Kyai Leadership: The Case of Pesantren Tebu Ireng)*. Malang: Kalimasahada Press.

Arto, Soegih. 1994. *Indonesia and I*. Singapore: Times Books International.

Babb, Lawrence. 1975. *The Divine Hierarchy: Popular Hinduism in Central India*. New York: Columbia University Press.

Barber, Benjamin. 1995. *Jihad vs. McWorld*. New York: Times Books.

Barth, Frederick. 1959. *Political Leadership among the Swat Pathans*. New York: Humanities Press, Inc.

Barton, Greg. 2002. *Abdurrahman Wahid: Muslim Democrat, Indonesian President*. Honolulu: University of Hawaii Press.

Bellah, Robert. 1968. "Meaning and Modernisation." *Religious Studies* 4: 37–45.

Benda, Harry J. 1958. *The Crescent and the Rising Sun: Indonesian Islam Under the Japanese Occupation*. The Hague: W. van Hoeve Ltd.

Boland, B.J. 1971. *The Struggle of Islam in Modern Indonesia*. The Hague: Martinus Nijhoff.

Bowen, John R. 1993a. "Discursive Monotheisms." *American Ethnologist* 20(1):185–190.

———. 1993b. *Muslims through Discourse: Religion and Ritual in Gayo Society*. Princeton, NJ: Princeton University Press.

Briggs, Charles L. 1996. "The Politics of Discursive Authority in Research on the Invention of Tradition." *Cultural Anthropology* 11(4):431–469.

Castles, Lance. 1966. "Note on the Islamic School at Gontor." *Indonesia* 1(1).

Cheneb, Mohammed ben. 1974. "Ibn Taimiya." In H.A.R. Gibb and J.H. Kramers, eds. *Shorter Encyclopedia of Islam*. Leiden, Netherlands: E.J. Brill, pp. 151–152.

Clement, Rob. 1987. "Editorial: Women in Development." *Pesantren's Linkage* 3(1):2.

Demerath, Peter. 1999. "The Cultural Production of Educational Utility in Pere Village, Papua New Guinea." *Comparative Education Review* 43(2): 162–192.

Dhofier, Zamakhsyari. 1980a. "Kinship and Marriage among the Javanese Kyai." *Indonesia* 29:47–58.

———. 1980b. *The Pesantren Tradition: A Study of the Role of the Kyai in the Maintenance of the Traditional Ideology of Islam in Java*. Doctoral Thesis submitted to The Australian National University.

———. 1982. *Tradisi Pesantren: Studi tentang Pandangan Hidup Kyai (The Pesantren Tradition: A Study of the Life View of Kyai)*. Jakarta: LP3ES.

———. 1995. "K.H Hasyim Asy'ari: Penggalang Islam Tradisional (Kyai Hasyim Aysari: Leader of Traditional Islam)." In Humaidy Abdussami and Ridwan Fakla, eds. *Lima Rais 'Am Nahdlatul Ulama*. Yogyakarta: Pustaka Pelajar.

———. 1999. *The Pesantren Tradition: A Study of the Role of the Kyai in the Maintenance of the Traditional Ideology of Islam in Java*. Tempe: Program for Southeast Asian Studies.

Eickelman, Dale. 1976. *Moroccan Islam: Tradition and Society in a Pilgrimage Center*. Austin: University of Texas Press.

———. 1982. "The Study of Islam in Local Contexts." *Contributions to Asian Studies* 17:1–18.

Eliade, Mircea. 1958. *Patterns in Comparative Religion*. Rosemary Sheed, trans. Lincoln: University of Nebraska Press.

el-Zein, Abdul Hamid. 1974. *The Sacred Meadows: A Structural Analysis of Religious Symbolism in an East African Town*. Evanston: Northwest University Press.

Federspiel, Howard. 1996. "The Endurance of Muslim Traditionalist Scholarship: An Analysis of the Writings of the Indonesian Scholar Siradjuddin Abbas." In Mark R. Woodward, ed. *Toward a New Paradigm: Recent Developments in Indonesian Islamic Thought*. Tempe: Program for Southeast Asian Studies.

Feillard, Andree. 1997. "Traditionalist Islam and the State in Indonesia: The Road to Legitimacy and the Renewal." In Robert Hefner and Patricia Horvatich, eds. *Islam in an Era of Nation-States*. Honolulu: University of Hawaii Press.

Florida, Nancy. 1995. *Writing the Past, Inscribing the Future: History as Prophecy in Colonial Java*. Durham: Duke University Press.

Frank, Andre Gunder. 1969. "The Development of Underdevelopment." In Andre Gunder Frank, ed. *Latin America: Development or Revolution*. New York: Monthly Review Press, pp. 3–17.

Friedman, Jonathon. 1992. "Narcissism, Roots and Postmodernity: The Constitution of Selfhood in the Global Crisis." In Scott Lash and Jonathon Friedman, eds. *Modernity and Identity*. Oxford: Blackwell, pp. 331–366.

Galba, Sindu. 1991. *Pesantren Sebagai Wasah Komunikasi (Pesantren as a Conduit of Communication)*. Jakarta: Departemen Pendikan dan Kebudayan.

Geertz, Clifford. 1960a. *Religion of Java*. Chicago: University of Chicago Press.

———. 1960b. "The Javanese Kijaji: The Changing Role of a Cultural Broker." *Comparative Studies in Society and History* 2(2):228–249.

———. 1968. *Islam Observed: Religious Development in Morocco and Indonesia*. Chicago: The University of Chicago Press.

George, Kenneth. 1996. *Showing Signs of Violence*. Berkeley: University of California Press.

Ghazali, Abu Hamid Muhammad. 1991. *Ihya Ulum-Id-Din*. Maulana Fazlul-Karim, trans. New Delhi: Islamic Book Services.

Ghofir, Abdul et al. 1982. *Sketsa Pondok Pesantren: Laporan Hasil Studi and Eksperimentasi Pondok Pesantren di Jawa Timur (Sketch of Pesantren: Research Report on East Javanese Pesantren)*. Malang: Fakultas Tarbiyah IAIN Sunan Ampel.

Gibb, Hamilton A.R. and Harold Bowen. 1958. *Islamic Society and the West*. London: Oxford University Press.

Giddens, Anthony. 1990. *The Consequences of Modernity*. Cambridge: Polity Press.

Gilsenan, M. 1982. *Recognizing Islam*. London: Croom Helm.

Goldziher, Ignaz. 1981 [1910]. *Introduction of Islamic Theology and Law.* Princeton, NJ: Princeton University Press.

Gramsci, Antonio. 1971. *Selections from the Prison Notebooks.* Geoffrey Smith and Quintin Hoare, trans. International Publisher Company.

Green, Arnold. 1978. *The Tunisan Ulama 1873–1915: Social Structure and Response to Ideological Currents.* Leiden, Netherlands: E.J. Brill.

Habermas, Jurgen. 1987. *The Philosophical Discourse of Modernity.* Cambridge: MIT Press.

Hamzah, K.H. Imron and Choirul Anam, eds. 1989. *Gus Dur Diadili Kiai-Kiai: Sebuah Dialog Mencari Kejelasan (Gus Dur is tried by Kyai: A Dialog Seeking Explanations).* Surabaya: Jawa Pos.

Hasan, Suhaib. 1994. *An Introduction to the Science of Hadith.* London: Al-Quran Society.

Hasyim, Yusuf. 1987. "The Role and Potential of Pesantren." *Pesantren's Linkage* 3(2):11–14.

Hefner, Robert. 1985. *Hindu Javanese.* Princeton, NJ: Princeton University Press.

———. 1987. "Islamizing Java? Religion and Politics in Rural East Java." *Journal of Asian Studies* 46:533–554.

———. 1996. "Islamizing Capitalism: On the Founding of Indonesia's First Islamic Bank." In Mark Woodward, ed. *Toward a New Paradigm: Recent Developments in Indonesian Islamic Thought.* Tempe: Arizona State University Program for Southeast Asian Studies.

———. 1997. "Islam and Democratization in Indonesia." In Robert Hefner and Patricia Horvatich, eds. *Islam in an Era of Nation-States.* Honolulu: University of Hawaii Press.

———. 2000. *Civil Islam: Muslims and Democratization in Indonesia.* Princeton, NJ: Princeton University Press.

Hobsbawm, Eric and Terrance Ranger, eds. 1983. *The Invention of Tradition.* Cambridge: Cambridge University Press.

Horikoshi, Hiroko. 1975. "The Dar ul-Islam Movement in West Java (1948–62)." *Indonesia* 20:59–86.

———. 1976. *A Traditional Leader in a Time of Change: The "Kijaji" and the "Ulamai" in West Java.* Urbana-Champaign: University of Illinois Doctoral Dissertation in Anthropology.

Howell, Julia Day. 2001. "Sufism and the Indonesian Islamic Revival." *Journal of Asian Studies* 60(3):701–730.

Hunter, James. 1991. *Culture Wars: The Struggle to Define America.* New York: Basic Books.

Huntington, Samuel P. 1966. "Political Modernization: America vs. Europe." *World Politics* 18(3):378–414.

———. 1993. "The Clash of Civilizations?" *Foreign Affairs* 72(3):22–49.

———. 1996. *The Clash of Civilizations and the Remaking of the World Order.* New York: Simon and Schuster.

Hye, M. Adbul. 1961. "Ash'arism." In M.M. Sharif, ed. *A History of Muslim Philosophy.* Delhi: Low Price Publications.

IAIN (Institut Agama Islam Negeri) Sunan Ampel Research Team. 1992. *Sistem Pendidikan Pesantren Kecil and Pengaruhnya Terhadap Perkembangan Kepribadian Anak (Educational System of Child Pesantren and the Influence on the Development of the Childrens' Sense of Personhood)*. Surabaya: Pusat Penelitian dan Pengabdian Pada Masyarakat, IAIN Sunan Ampel.

Jones, Sidney. 1991. "The Javanese Pesantren: Between Elite and Peasantry." In Charles F. Keyes, ed. *Reshaping Local Worlds: Formal Education and Cultural Change in Rural Southeast Asia*. New Haven: Yale Center for International and Area Studies—Southeast Asia Studies.

Kafanjanji, A.R. n.d. *Menyinkap Kisah Keteladanan Perjuangan Walisongo (The Story of the Struggle of the Wali Songo)*. Surabaya: Anugerah.

Kahin, George McT. 1952. *Nationalism and Revolution in Indonesia*. Ithaca, NY: Cornell University Press.

Kamal, Musthafa, Chusnon Yusuf, and Rosyeh Sholeh. 1988. *Muhammadiyah Sebagai Gerakan Islam (Muhammadiyah as an Islamic Movement)*. Yogyakarta: Penerbit Persatuan. Sixth printing.

Kantor Statistik Propinsi Jawa Timur (East Java Statistics Office). 1993. *Jawa Timur Dalam Angka (East Java Figures)*. Surabaya: Kantor Statistik Propinsi Jawa Timur.

Kartodirdjo, Sartono. 1966. *The Peasants' Revolt of Banten in 1888*. The Hague: Martinus Nijhoff.

Keddie, Nikki R., ed. 1972. *Scholar, Saints, and Sufis: Muslim Religious Institutions in the Middle East since 1500*. Berkeley and Los Angeles: University of California Press.

Keeler, Ward. 1987. *Javanese Shadow Plays, Javanese Selves*. Princeton, NJ: Princeton University Press.

Kern, R.A. 1953. "Pasantren." In H.A.R. Gibb and J.H. Kramers, eds. *Shorter Encyclopedia of Islam*, pp. 460–462.

Koentjaraningrat, R.M. 1957. *A Preliminary Description of the Javanese Kinship System*. New Haven, CT: Yale University Southeast Asia Studies Cultural Report Series.

———. 1963. "Review of *The Religion of Java*." *Masalah Ilmu-Ilmu Sastra Indonesia* No. 2; 188–191.

Kumar, Ann. 1985. *The Diary of a Javanese Muslim: Religion, Politics, and the Pesantren 1833–1886*. Canberra: Australia National University Faculty of Asian Studies.

Lane, Edward William. 1860. *Manners and Customs of the Modern Egyptians*. London: John Murray.

Lansing, J. Stephen. 1983. *The Three Worlds of Bali*. New York: Praeger.

———. 1995. *The Balinese*. New York, NY: Harcourt Brace Publishers.

Lapidus, Ira M. 1988. *A History of Islamic Societies*. New York: Cambridge University Press.

Lawrence, Bruce. 1989. *Defenders of God*. San Francisco: Harper and Row, Publishers.

Leacock, Eleanor Burke. 1976. "Education in Africa: Myths of Modernization." In Calhoun, Craig and Francis Ianni, eds. *The Anthropological Study of Education*. The Hague: Mouton Publishers, pp. 239–250.

Levine, Lawrence. 1996. *The Opening of the American Mind: Canons, Culture, and History*. Boston: Beacon Press.

Lewis, Bernard. 1997. "The West and the Middle East." *Foreign Affairs* 76(1):114–130.

Li, Defeng. 1999. "Modernization and Teacher Education in China." *Teaching and Teacher Education* 15(2): 179–192.

Lukens-Bull, Ronald A. 1996. "Metaphorical Aspects of Indonesian Islamic Discourse About Development." In Mark Woodward, ed. *Toward a New Paradigm: Recent Developments in Indonesian Islamic Thought*. Tempe: Arizona State University Program for Southeast Asian Studies.

Lyon, Margo L. 1970. *Bases of Conflict in Rural Java*. Berkeley: University of California, Berkeley Center for South and Southeast Asia Studies Research Monograph Series.

Madjid, Nurcholish. 1996. "In Search of Islamic Roots for Modern Pluralism: The Indonesian Experiences." In Mark Woodward, ed. *Toward a New Paradigm: Recent Developments in Indonesian Islamic Thought*. Tempe: Arizona State University Program for Southeast Asian Studies.

Makdisi, George. 1981. *The Rise of Colleges: Institutions of Learning in Islam and the West*. Edinburgh: Edinburgh University Press.

Mansurnoor, Iik Arifin. 1990. *Islam in an Indonesian World: Ulama of Madura*. Yogyakarta: Gadjah Mada University Press.

Mater, Nadire. 1996. "Turkey: High Priced Private Schools No Answer to Education Crisis." InterPress Service English News Wire, 10-09-1996.

———. 1997. "Turkey: Religious Seminaries the Next Battleground for the Army." InterPress Service English News Wire, 03-12-1997.

Metcalf, Barbara. 1982. *Islamic Revival in British India: Beoban 1860–1900*. Princeton, NJ: Princeton University Press.

Mitchell, Edna. 1976. "The New Education Plan in Nepal: Balancing Conflicting Values for National Survival." In Calhoun, Craig, and Francis Ianni, eds. *The Anthropological Study of Education*. The Hague: Mouton Publishers, pp. 159–170.

Moertono, Soemarsaid. 1968. *State and Statecraft in Old Java. A Study of the Later Mataram Period, 16th to 19th Century*. Ithaca, NY: Cornell University.

Mottahedeh, Roy. 1985. *The Mantle of the Prophet: Religion and Politics in Iran*. New York: Pantheon Books.

Muhaimin, Abdul Ghoffur. 1995. *The Islamic Traditions of Cirebon: Ibadat and Adat among Javanese Muslim*. Unpublished Doctoral Dissertation, Australia National University.

Nasution, Harun and the Editorial Team from IAIN Sarif Hidayatullah. 1992. *Ensiklopedi Islam Indonesia*. Jakarta: Djambatan Anggotta IKAPI.

Noer, Deliar. 1973. *The Modernists Muslim Movement in Indonesian 1900–1942*. London: Oxford University Press.

————. 1978. *Administration of Islam in Indonesia.* Ithaca, NY: Cornell Modern Indonesia Project.

Okello Abungu, George H. 1994. "Islam on the Kenyan Coast: An Overview of Kenyan Coastal Sacred Sites." In David Carmichael, Jane Hubert, Brian Reeves, and Audhild Schanche, eds. *Sacred Sites, Sacred Places.* London: Routledge.

Peacock, James. 1978. *Muslim Puritans: Reformist Psychology in Southeast Asian Islam.* Berkeley: University of California Press.

Pederson, J. 1953a. "Masdjid." In H.A.R. Gibb and J.H. Kramers, eds. *Shorter Encyclopedia of Islam.* Leiden, Netherlands: E.J. Brill, pp. 330–353.

————. 1953b. "Madrasa." In H.A.R. Gibb and J.H. Kramers, eds. *Shorter Encyclopedia of Islam.* Leiden, Netherlands: E.J. Brill, pp. 300–310.

Pemberton, John. 1994. *On the Subject of "Java."* Ithaca, NY: Cornell University Press.

Pipes, Daniel. 1983. *In the Path of God: Islam and Political Power.* New York: Basic Books.

Pranowo, Bambang. 1991. "Traditional Islam in Contemporary Rural Java." In M.C. Ricklefs, ed. *Islam in the Indonesian Social Context.* Clayton: Monash University Center for Southeast Asian Studies Lecture Series.

Prasodjo, Sudjoko, M. Zamroni, M. Mastuhu, Sardjono Geonari, Nurcholish Madjid, and M. Dawam Rahardo. 1974. *Profil Pesantren: Laporan Hasil Penelitian Pesantren Al-Falak dan Delapan Pesantren Lain di Bogor. (Pesantren Profile: A Research Report on Pesantren Al-Falak and Eight other Pesantren in Bogor).* Jakarta: LP3ES.

Rachman, Abdul. 1997. *The Pesantren Architects and their Socio-Religious Teachings (1850–1950).* Unpublished Ph.D. Dissertation. Los Angeles: University of California.

Rahman, Fazlur. 1979. *Islam, Second Edition.* Chicago: University of Chicago Press.

Reichmuth, Stefan. 1993. "Islamic Learning and 'Western' Education in Ilorin." In Louis Brenner, ed. *Muslim Identity and Social Change in Sub-Saharan Africa.* London: Hurst and Company.

Ricklefs, M.C. 1979. "Six Centuries of Islamization in Java." In N. Levtzion, ed. *Conversion to Islam.* New York: Holmes and Meier.

Schimmel, Annemarie. 1991. "Sacred Geography in Islam." In Jamie Scott and Paul Simpson-Housely, eds. *Sacred Places and Profane Spaces: Essays in the Geographics of Judaism, Christianity, and Islam.* New York: Greenwood Press.

Schulz, Dorothea. 1997. "Praise without Enchantment: Griots, Broadcast Media, and the Politics of Tradition in Mali." *Africa Today* 44(4): 443–465.

Schwarz, Adam. 1994. *A Nation in Waiting: Indonesia in the 1990s.* St. Leonards: Allen and Unwin.

Scott, James C. 1990. *Domination and the Arts of Resistance: Hidden Transcripts.* New Haven: Yale University Press.

Sheik, M. Saeed. 1961. "Al-Ghazali—Mysticism." In M.M. Sharif, ed. *A History of Muslim Philosophy.* Delhi: Low Price Publications.

Singh, Hari. 1998. "Tradition, UMNO and Political Succession in Malaysia." *Third World Quarterly* 19(2):241–255.

Sperling, David. 1993. "Rural Madrasas of the Southern Kenya Coast, 1971–1992." In Louis Brenner, ed. *Muslim Identity and Social Change in Sub-Saharan Africa.* London: Hurst and Company.

Stange, Paul. 1990. "Javanism as Text or Praxis." *Anthropological Forum* 6(2):237–255.

Steenbrink, Karel. 1974. *Pesantren, Madrasah, Sekoloh: Pendidikan Islam dalam Kurun Moderen (Pesantren, Madrasah, Schools: Islamic Education in the Modern Era).* Jakarta: LP3ES.

Syihab, Muhammad Asad. 1994. *Hadratussyaikh Muhammad Hasyim Asy'arie: Perintis Kemerdekaan Indonesia.* KHA Bisri, Mustofa, trans. Yogyakarta: Titian Ilahi Press.

Syukri, M. 1994. *Pola Pembelajaran Ketrampilan Para Santri Sebagai Pengembangan Pendidikan Luar Sekolah di Pondok Pesantren: Studi Kasus Pondok Pesantren An-Nur II Al-Murtadlo Bululawang (Skills Training for Santri as the Development of Extra-Curricular Education in Pesantren: Case Study An Nur II Al-Murtadlo Bululawang).* MA level thesis. Institut Keguran dan Ilmu Pendidikan: Malang.

Turner, Victor. 1969. *The Ritual Process: Structure and Anti-Structure.* New York: Aldine de Gruyter.

Tsing, Anna. 1993. *In the Realm of the Diamond Queen.* Princeton, NJ: Princeton University Press.

van Bruinessen, Martin. 1990. "Kitab Kuning: Books in Arabic Script Used in the Pesantren Milieu." *Bijdragen* 146(2/3):226–269.

———. 2002. "Back to Situbondo? Nahdlatul Ulama Attitudes towards Abdurrahman Wahid's Presidency and His Fall." In Henk Schulte Nordholt and Irwan Abdullah, eds. *Indonesia: In Search of Transition.* Yogyakarta: Pustaka Pelajar, pp. 15–46.

Vikør, Knut S. 1995. "The Development of Ijtihad and Islamic Reform, 1750–1850." Paper presented at the Third Nordic Conference on Middle Eastern Studies: Ethnic Encounter and Culture Change, Joensuu, Finland 19–22 June 1995. <http://www.hf-fak.uib.no/smi/paj/Vikor.html>.

Ward, Robert E. and Dankwart A. Rustow. 1964. *Political Modernization in Japan and Turkey.* Princeton, NJ: Princeton University Press.

Watt, William Montgomery. 1974. *The Majesty That Was Islam.* New York: Praeger.

Weatherbee, Donald E. 1985. "Indonesia in 1984: Pancasila, Politics, and Power." *Asian Survey* 25(2):187–197.

White, Benjamin. 1983. " 'Agricultural Involution' and its Critics: Twenty Years After." *Bulletin of Concerned Asian Scholars* 15(2):18–31.

Wood, Hugh. 1976. "Agents of Education and Development in Nepal." In Calhoun, Craig and Francis Ianni, eds. *The Anthropological Study of Education.* The Hague: Mouton Publishers, pp. 146–170.

Woodward, Mark. 1988. "The Slametan: Textual Knowledge and Ritual Performance in Central Javanese Islam." *History of Religions* 28(1):54–89.

————. 1989. *Islam in Java: Normative Piety and Mysticism in the Sultanate of Yogyakarta*. Tucson: University of Arizona Press.

————. 1996a. "Talking Across Paradigms: Indonesia, Islam, and Orientalism." In Mark Woodward, ed. *Toward a New Paradigm: Recent Developments in Indonesian Islamic Thought*. Tempe: Arizona State University Program for Southeast Asian Studies.

————. 1996b. "Conversations with Abdurrahman Wahid." In Mark Woodward, ed. *Toward a New Paradigm: Recent Developments in Indonesian Islamic Thought*. Tempe: Arizona State University Program for Southeast Asian Studies.

Yunus, Mahmud. 1979. *Sejarah Pendidikan Islam in Indonesia (History of Islamic Education in Indonesia)*. Jakarta: Mutiara.

Zaini, A. Wahid. 1994. *Dunia Pemikiran Kaum Santri*. Yogyakarta: LKPSM.

Index